Sustainability Certification Program for Game Changers

Learn. Practice. Get Certified.

ANGEL BERNIZ

International Institute of Executive Careers (IIEC)

Program Contents

Welcome to the Program

Thank you very much for purchasing this book.

First of all, let's connect on LinkedIn and join my **Game Changers** Newsletter on LinkedIn, so that you can get my insights, learnings, and exclusive contents to help you get career success:

https://www.linkedin.com/in/angelberniz

I would love you to share your experience with me after completing the program. You can message me via LinkedIn.

Okay, let's move on!

Certification Program

This book is the foundation for the **IIEC Sustainability Certification Program™** available here:

https://thebestmba.org/program/sustainability-certification

The Certification Program is available in the online program and will improve your job opportunities and certify your skills gained from this book.

The program exams included in the online program are based on the exam questions covered along this book. And once you sign up, you can take your online program whenever you want (no deadline limits). Just sign up today for your Certification Program and certify your skills.

Now, let's start enjoying this journey.

Preface

Welcome to this book on Sustainability, a program designed to inspire learners, corporate leaders, and organizations to take action and become game changers. As you turn these pages, you will learn about the challenges we face as a global community and the opportunities that exist to create a sustainable future for all, including businesses.

The world is facing complex and interconnected challenges that threaten the well-being of people and the planet, and corporations have a significant role to play in addressing them. Climate change, resource depletion, biodiversity loss, and social inequality are among the most pressing issues that require urgent action from the private sector.

This book is not just a collection of ideas or theories; it is a practical guide that will help you transform your business and the world around you. Each module is structured around a specific theme, and includes practical examples, case studies, and action plans that will empower you to make a meaningful impact.

The goal of this book is not to make you feel guilty or overwhelmed by the scale of the challenges we face. Instead, it is to motivate you to take action and become a game changer. You don't have to be an expert or a hero to make a difference. All you need is the willingness to learn, the courage to act, and the persistence to overcome obstacles.

Yes, that sentence is grammatically correct and it conveys the idea that sustainability is not only about taking responsibility towards the planet, but also towards society and future generations. Furthermore, adopting sustainable practices can create long-term value for your business. Adopting sustainable practices will allow you to reduce costs, mitigate risks, enhance your brand reputation, and attract and retain customers and talent.

I encourage you to embrace this journey with an open mind and a positive attitude. Let this book be your guide, but also your inspiration to go beyond the pages and make a real difference in your business and your community. Together, we can create a more sustainable, resilient, and prosperous future for all. Let's start today!

Acknowledgments

My notes at the university were: "Early adopters are the wealthiest people on earth."

Little did I know how those words would shape my path. As I reflect on my journey, I am overwhelmed with gratitude for the countless individuals who accompanied, guided, and inspired me along the way.

I owe a profound debt of gratitude to my mentors, colleagues, and collaborators in executive education and the technology industry. Their unwavering support and willingness to share their knowledge and perspectives have been invaluable. Their guidance has shaped my career and fueled my passion.

To my students at the International Institute of Executive Careers (IIEC), you have been my constant inspiration. Your curiosity and dedication have made me a better educator and challenged me to refine my ideas. Your insights have significantly contributed to this book.

I also acknowledge the remarkable researchers, professionals, and practitioners who have paved the way for our understanding of the potential of emerging technologies. Their visionary thinking and tireless pursuit of knowledge continue to inspire individuals worldwide, including myself.

I am deeply grateful for the unwavering love, support, and understanding of my friends and family. Your belief in me and willingness to engage with my ideas have given me strength and resilience. This book is dedicated to you with love and gratitude.

As I conclude, I am reminded of the interconnected relationships, experiences, and ideas that have shaped this book. To all who have touched my life, big or small, thank you. Your presence has been truly meaningful. Together, let us continue exploring the evolving landscape of technology, guided by the belief that early adopters can shape the future.

About the Author

Angel Berniz is a seasoned Business and Technology executive, strategist, and advisor with over 20 years of experience in the technology industry. He has held various leadership positions in several companies and has worked with clients in different industries, helping them navigate cloud transformation, innovation, sustainability, emerging technologies, and cybersecurity.

Angel Berniz is also a recognized keynote speaker and author on topics related to business technologies, leadership, and innovation. He has been featured in several conferences, seminars, and publications. He is also a professor and lecturer in international executive programs of several Business Schools.

You can explore Angel Berniz's extensive collection of books available on Amazon:

https://www.amazon.com/author/angelberniz

Module 1. Introduction to Sustainability

Introduction to Sustainability

Welcome to Module 1 on Introduction to Sustainability. In this module, you will embark on a journey to explore the world of Environmental, Social, and Governance (ESG) principles and their vital role in achieving a sustainable future for businesses and society. As you progress through this module, you will uncover the significance of sustainability in business, learn about the various ESG criteria, and discover the essential Key Performance Indicators (KPIs) for effective ESG performance measurement.

Moreover, you will delve into the world of technology for sustainability, where you'll understand how innovations in clean energy, circular economy, and remote work can contribute to a more sustainable business ecosystem. With practical examples and engaging content, you'll develop a comprehensive understanding of how ESG can create value, mitigate risks, attract investors, enhance customer loyalty, and foster employee engagement.

By the end of this module, you will have a solid foundation in ESG and sustainability concepts, enabling you to make informed decisions and implement effective strategies in your business. So, get ready to dive into this fascinating world of sustainable business practices and discover how you can make a meaningful impact on our planet and society!

Understanding the ESG Concept

In today's rapidly evolving global landscape, organizations are increasingly recognizing the importance of sustainability, both as a strategic imperative and a tool for addressing pressing societal and environmental challenges. The Environmental, Social, and Governance (ESG) framework has emerged as a vital approach for companies to assess, measure, and report their sustainability performance. By incorporating ESG factors into their decision-making processes, businesses can not only mitigate risks but also capitalize on new opportunities to drive long-term value creation.

ESG is an umbrella term encompassing a range of non-financial factors that investors and stakeholders consider to evaluate a company's ethical and sustainable practices. These factors are typically classified into three main categories: Environmental, Social, and Governance.

Environmental factors relate to a company's impact on the natural environment, including its efforts to mitigate climate change, reduce greenhouse gas emissions, conserve resources, and minimize waste and pollution. For instance, a manufacturing firm that adopts energy-efficient technologies, uses recycled materials, and limits its emissions of hazardous substances demonstrates its commitment to environmental sustainability.

Social factors encompass the relationships that a company maintains with its employees, suppliers, customers, and the communities in which it operates. These factors include labor rights, health and safety, diversity and inclusion, and community engagement. For example, a retail company that implements fair labor practices, promotes a diverse workforce, and supports local community development projects exhibits a strong focus on social sustainability.

Governance factors pertain to the internal systems and structures that guide a company's decision-making, ethical conduct, and overall management. These factors include board composition and diversity, executive compensation, shareholder rights, and adherence to business ethics and regulatory compliance. A financial institution that upholds robust corporate governance, maintains transparency in its operations, and actively addresses corruption risks exemplifies good governance.

In recent years, ESG integration has gained significant traction, as organizations worldwide recognize the value of sustainable practices. This shift is driven by several factors, including mounting regulatory pressures, growing investor demand for responsible investment, and the increasing awareness of the link between ESG performance and long-term financial performance.

For example, a study conducted by the New York University Stern Center for Sustainable Business found that firms with strong ESG performance demonstrated higher profitability, lower volatility, and reduced downside risk compared to their low ESG-performing counterparts. This empirical evidence reinforces the business case for embedding ESG principles into corporate strategy and operations.

In summary, sustainability serves as a comprehensive framework for companies to evaluate and address the environmental, social, and governance aspects of their business. By understanding and managing their ESG performance, organizations can foster a more resilient, sustainable, and competitive future in an increasingly complex world.

Importance of Sustainability in Business

Incorporating sustainability into your business strategy is essential for several reasons, as it can create value, reduce risks, and enhance your company's reputation. By understanding the significance of ESG factors, you can unlock new opportunities and drive long-term success for your business. Let's explore the key reasons why sustainability matters.

- **Value creation:** Aligning your business with ESG principles can lead to cost savings, improved efficiency, and enhanced innovation. For example, by investing in renewable energy sources, you can reduce your energy costs, while adopting circular economy practices can minimize waste and increase resource efficiency.

- **Risk mitigation:** Addressing ESG factors can help you identify and manage potential risks, such as regulatory changes, reputational damage, or supply chain disruptions. For instance, by ensuring fair labor practices, you can avoid potential lawsuits and maintain a positive brand image.

- **Investor appeal:** Investors are increasingly considering ESG performance when making investment decisions. By demonstrating your commitment to sustainability, you can attract capital from socially responsible investors and benefit from favorable financing terms.

- **Customer loyalty:** Consumers are becoming more conscious of the environmental and social impact of the products and services they purchase. By integrating ESG principles into your operations, you can meet the growing demand for sustainable offerings and strengthen customer loyalty.

- **Employee engagement:** Embracing ESG values can help you attract and retain top talent, as employees are more likely to feel motivated and engaged in companies that prioritize sustainability. For example, promoting diversity and inclusion can create a more inclusive work environment, leading to higher employee satisfaction and retention.
- **Competitive advantage:** By adopting ESG best practices, you can differentiate your company from competitors and enhance your market position. For instance, implementing cutting-edge sustainable technologies or obtaining sustainability certifications can distinguish your brand and drive consumer preference.

In conclusion, prioritizing sustainability in your business is crucial for long-term success. By embracing ESG principles, you can create value, mitigate risks, attract investment, foster customer loyalty, engage employees, and gain a competitive edge. By doing so, you can build a more resilient and sustainable future for your business and stakeholders.

Practical examples of value creation

Value creation through sustainability can take many forms, as companies invest in initiatives that not only benefit the environment and society but also generate financial returns. Here are some practical examples:

- **Energy efficiency:** A manufacturing company invests in energy-efficient equipment and technologies, such as LED lighting and energy management systems. This reduces energy consumption and lowers utility costs, resulting in significant cost savings over time.
- **Waste reduction:** A food and beverage company implements a waste reduction program by optimizing packaging materials, reducing food waste, and recycling or composting leftover materials. This leads to cost savings in

terms of raw materials, waste disposal, and packaging expenses.

- **Water conservation:** A textile factory implements water-saving measures, such as using water-efficient dyeing processes and recycling wastewater for reuse in its operations. This reduces water consumption, lowers water treatment costs, and minimizes the risk of water scarcity-related disruptions.
- **Sustainable sourcing:** A consumer goods company commits to sourcing raw materials from certified sustainable suppliers, reducing the risk of supply chain disruptions and potentially securing better pricing due to increased demand for responsibly sourced products.
- **Product innovation:** An electronics firm develops eco-friendly products, such as energy-efficient appliances or devices made from recycled materials. These products cater to the growing consumer demand for sustainable options, resulting in increased sales and market share.
- **Circular economy practices:** A furniture company adopts a circular business model by offering customers the option to return used furniture for refurbishment or recycling in exchange for a discount on new purchases. This approach reduces waste, generates additional revenue from refurbished products, and strengthens customer loyalty.

These examples demonstrate how incorporating sustainability into business operations can create value by reducing costs, driving innovation, and enhancing overall performance. By prioritizing ESG principles, companies can unlock new opportunities and contribute to long-term success.

Practical examples of risk mitigation

Risk mitigation through sustainability involves addressing environmental, social, and governance factors to reduce potential threats to your business. Here are some practical examples:

- **Climate change adaptation:** A real estate company invests in climate-resilient infrastructure, such as green roofs and flood-resistant designs, to protect its properties from extreme weather events. This reduces the risk of property damage, insurance claims, and business disruptions.
- **Supply chain management:** A clothing retailer conducts regular audits of its suppliers to ensure compliance with labor and environmental standards. By identifying and addressing potential issues early, the company minimizes the risk of reputational damage, supply chain disruptions, and regulatory fines.
- **Regulatory compliance:** A chemical company closely monitors changes in environmental regulations and proactively adjusts its operations to meet new standards. This mitigates the risk of non-compliance, legal penalties, and negative publicity.
- **Data privacy and cybersecurity:** A financial services firm implements robust data privacy policies and invests in state-of-the-art cybersecurity measures to protect customer information. This reduces the risk of data breaches, loss of customer trust, and potential legal liabilities.
- **Ethical governance:** A pharmaceutical company establishes a strong code of ethics, transparent reporting practices, and an independent board of directors to ensure sound decision-making and prevent corruption. This minimizes the risk of financial scandals and reputational harm.

- **Health and safety:** A construction company invests in employee training, safety equipment, and robust safety protocols to minimize workplace accidents. This reduces the risk of employee injuries, lost productivity, and potential legal claims.

Proactively addressing ESG risks will allow companies to protect their assets, maintain their reputation, and ensure the long-term viability of their operations. Implementing ESG risk mitigation strategies not only benefits the environment and society, but also contributes to the overall resilience and success of the business.

Practical examples of investor appeal

Investor appeal through sustainability involves demonstrating a strong commitment to environmental, social, and governance factors, which can attract the attention of socially responsible investors. Here are some practical examples:

- **Transparent reporting:** A tech company adopts transparent ESG reporting practices by publishing annual sustainability reports that detail its environmental and social performance, as well as governance policies. This transparency allows investors to assess the company's ESG performance and attracts those who prioritize responsible investing.

- **Carbon neutrality commitment:** An energy company sets ambitious targets to achieve carbon neutrality by investing in renewable energy, implementing energy efficiency measures, and participating in carbon offset projects. This commitment appeals to investors who focus on companies taking action against climate change.

- **Diversity and inclusion initiatives:** A financial services firm promotes diversity and inclusion in its workforce and leadership by implementing fair hiring practices, offering training programs, and setting diversity targets. Investors who value social responsibility find this commitment appealing, as diverse and inclusive companies often perform better.

- **Sustainable product offerings:** A consumer goods company launches a range of eco-friendly products made from recycled materials or with reduced environmental impact. This product line attracts investors who seek companies with a positive impact on society and the environment.

- **Community engagement:** A retail company actively engages with the communities in which it operates, supporting local initiatives and investing in community development projects. This social commitment appeals to investors who prioritize companies that contribute positively to the well-being of society.

- **ESG-focused indices and funds:** Companies with strong ESG performance are often included in ESG-focused indices and funds, such as the Dow Jones Sustainability Index or the MSCI ESG Leaders Index. Inclusion in these indices not only validates a company's ESG commitment but also increases its visibility among investors seeking responsible investment options.

Demonstrating a commitment to Sustainability will allow companies to attract the interest of a growing number of investors who prioritize responsible investing. This can result in increased access to capital, more favorable financing terms, and a higher valuation for the company, ultimately contributing to long-term business success.

Practical examples of customer loyalty

Customer loyalty through Sustainability involves meeting the growing demand for environmentally and socially responsible products and services, which can lead to stronger relationships with customers. Here are some practical examples:

- **Eco-friendly packaging:** A food company switches to recyclable or biodegradable packaging materials, reducing its environmental footprint and appealing to environmentally conscious consumers who appreciate and support sustainable practices.

- **Fair labor practices:** A clothing brand ensures that its garments are produced under fair labor conditions, partnering with suppliers who pay living wages and maintain safe working environments. Customers who prioritize ethical production are more likely to remain loyal to brands that share their values.

- **Organic and locally sourced products:** A restaurant chain commits to using organic, locally sourced ingredients in its menu offerings, supporting local farmers and reducing the carbon footprint associated with transportation. Customers who value sustainable food choices are more likely to become repeat patrons.

- **Transparency and traceability:** A coffee company implements a traceability system that allows customers to track the origin of their coffee beans and learn about the environmental and social practices employed throughout the supply chain. This transparency can build trust and loyalty among customers who appreciate knowing the story behind their purchases.

- **Corporate social responsibility (CSR) initiatives:** A telecommunications company launches a CSR program that includes initiatives such as providing free internet access to underserved communities or donating a portion of profits to environmental causes. Customers who value social responsibility are more likely to remain loyal to companies that contribute to the greater good.

- **Responsiveness to customer feedback:** A beauty products company actively solicits customer feedback on its sustainability efforts and incorporates suggestions into its product development and business practices. Demonstrating a genuine commitment to customer input can strengthen loyalty and foster a sense of shared values.

Integrating ESG principles into their products and services will allow companies to meet the growing consumer demand for sustainable options and strengthen customer loyalty. By doing so, businesses can enhance their reputation, increase customer retention, and drive long-term success.

Practical examples of employee engagement

Employee engagement through sustainability involves creating a work environment that fosters a sense of purpose, motivation, and commitment among employees. Here are some practical examples:

- **Inclusive workplace culture:** A technology firm promotes diversity and inclusion by offering sensitivity training, employee resource groups, and mentorship programs. By fostering an inclusive culture, employees feel valued and respected, leading to higher engagement and retention.

- **Training and development opportunities:** A retail company invests in employee training and development, providing opportunities to enhance skills and knowledge in sustainability and ESG topics. This commitment to professional growth can increase employee motivation and engagement.

- **Employee volunteering and giving programs:** A healthcare organization establishes employee volunteering and giving programs, encouraging staff to participate in community service activities and matching employee donations to charitable causes. These initiatives can boost employee morale and strengthen their connection to the company's values.

- **Sustainable workplace practices:** An advertising agency implements sustainable workplace practices, such as recycling programs, energy-efficient lighting, and remote work options to reduce its environmental impact. Employees who are passionate about sustainability are more likely to feel engaged and proud of their company's efforts.

- **Employee wellness and work-life balance:** A financial services company prioritizes employee wellness and work-life balance by offering flexible working hours, mental health resources, and wellness programs. Supporting employee well-being can lead to higher engagement and job satisfaction.

- **Recognition and rewards for ESG achievements:** A manufacturing company recognizes and rewards employees for their contributions to sustainability initiatives, such as reducing energy consumption or developing innovative, eco-friendly products. This recognition can motivate employees to continue pursuing ESG-related goals and foster a sense of pride in their work.

Embracing ESG values and creating a work environment that supports employee engagement will allow companies to attract and retain top talent, increase job satisfaction, and improve overall performance. Engaged employees are more likely to be productive, innovative, and dedicated to the company's long-term success.

Practical examples of competitive advantage

Competitive advantage through sustainability involves adopting responsible business practices that differentiate your company from competitors, leading to improved market positioning and long-term success. Here are some practical examples:

- **Innovation and product differentiation:** An automotive company invests in the research and development of electric vehicles and charging infrastructure, allowing it to stay ahead of competitors and cater to the growing demand for sustainable transportation options.

- **Supply chain resilience:** A consumer electronics firm implements rigorous supplier assessments and monitoring to ensure adherence to environmental and social standards, reducing the risk of supply chain disruptions and enhancing its reputation among consumers and investors.
- **Operational efficiency:** A logistics company adopts energy-efficient technologies and practices, such as route optimization and fuel-efficient vehicles, resulting in cost savings and reduced environmental impact. This efficiency can set the company apart from competitors and attract eco-conscious customers.
- **Brand reputation and trust:** A cosmetics company commits to cruelty-free and environmentally friendly production practices, gaining the trust of customers who prioritize ethical consumption. This strong brand reputation can lead to increased customer loyalty and market share.
- **Access to capital and financing:** A renewable energy company demonstrates strong ESG performance and attracts the interest of impact investors, resulting in more favorable financing terms and a broader pool of potential investors.
- **Regulatory compliance and adaptability:** A chemical manufacturer proactively monitors and adapts to evolving environmental regulations, positioning itself as an industry leader and reducing the risk of fines and penalties associated with non-compliance.

Incorporating sustainability into their operations and strategies will allow companies to gain a competitive edge that sets them apart from their rivals. This advantage can result in increased market share, customer loyalty, and long-term success, as businesses become better equipped to navigate the challenges and opportunities presented by an increasingly sustainability-conscious world.

What is Greenwashing and How to Avoid it

As the world becomes more environmentally conscious, companies are eager to capitalize on this trend by marketing themselves as eco-friendly. However, not all companies are actually committed to sustainability, and some engage in a practice known as "greenwashing." Greenwashing refers to the act of making a brand appear more environmentally friendly than it actually is. This can take many forms, from deceptive marketing gimmicks to simply changing the packaging of an established product.

One of the most common forms of greenwashing is "**green-crowding**." This occurs when a company has a low profile within a group of companies that are signatories of global alliances, allowing them to rely on safety in numbers and move at the speed of the slowest adopter of sustainability policies without being spotted. This makes it difficult for consumers to distinguish truly sustainable brands from those that are merely trying to appear green.

Another tactic is "**green-lighting**." This occurs when companies spotlight a particularly green feature of their operations or products, drawing attention away from environmentally damaging activities being conducted elsewhere, either in their production or supply chain. This misleads consumers into thinking that the company is more environmentally friendly than it actually is.

"**Green-labelling**" is another common tactic used in greenwashing. Marketers call something green or sustainable, but a closer examination reveals this to be misleading. Products are promoted in green packaging, with plant symbols or deceptive wording, causing consumers to infer their eco credentials. This is particularly common in industries where there are no clear regulations around sustainability claims, such as the fashion industry.

Another tactic is "**green-hushing**." This occurs when organizations deliberately choose to under-report or hide their green or ESG (environmental, social, and governance) credentials to evade investor and public scrutiny. By remaining vague, they may be giving the impression that they are greener than they actually are.

"**Green-rinsing**" is when a company regularly changes, reduces, or delays its ESG targets before they are achieved. This makes it difficult to assess the company's true commitment to sustainability, as they are constantly shifting their goals.

Finally, there is "**green-shifting**." This occurs when companies imply in their marketing campaigns that the consumer is at fault and shift the blame on to them. For example, a company may promote a product as being sustainable, but then blame the consumer for not using it in an environmentally friendly way.

Guidance on how to avoid greenwashing

Here are some tips for avoiding greenwashing and making informed choices about sustainable products and companies:

1. **Do your research:** Look beyond the marketing claims and do some research on the company's sustainability practices. Check their website, annual sustainability reports, and third-party certifications or ratings.

2. **Check for certifications:** Look for recognized third-party certifications like the Forest Stewardship Council (FSC), the Rainforest Alliance, or the Global Organic Textile Standard (GOTS). These certifications indicate that the product has met specific environmental and social standards.

3. **Read labels and packaging:** Pay attention to the details of the product, like its ingredients or materials, and avoid products with vague or misleading claims.

4. **Compare products:** Compare the sustainability credentials of different products in the same category. This can help you identify which products are truly sustainable and which are simply greenwashing.

5. **Look beyond the product:** Consider the company's overall sustainability practices, including its supply chain, energy use, and waste reduction efforts.

6. **Follow sustainability news:** Stay informed about current sustainability issues and trends, and read up on the latest research and innovations in sustainable products and practices.

Following these tips and doing your research will allow you to make informed choices about the products you buy and support companies that are truly committed to sustainability.

In conclusion, greenwashing is a deceptive practice that undermines the efforts of truly sustainable companies and misleads consumers. As consumers become more aware of greenwashing tactics, it is important for companies to be transparent about their sustainability credentials and avoid misleading marketing strategies. Only then can consumers make informed choices about the products they buy and the companies they support.

ESG Criteria and Key Performance Indicators (KPIs)

To effectively measure and evaluate your company's performance in sustainability, it's essential to understand the ESG criteria and establish relevant Key Performance Indicators (KPIs). These criteria and KPIs help you track progress, identify areas for improvement, and make informed decisions about your sustainability initiatives.

Environmental Criteria

These criteria focus on your company's impact on the environment, including energy consumption, greenhouse gas emissions, waste management, and resource conservation. Some practical examples of environmental KPIs include:

- **Carbon footprint:** Measure the total greenhouse gas emissions produced by your company, either in absolute terms or per unit of production.

- **Energy efficiency:** Track the amount of energy used per unit of output or revenue to gauge your company's energy efficiency.

- **Waste reduction:** Monitor the amount of waste generated, recycled, or diverted from landfills to assess your waste management practices.

Here are some practical examples of Key Performance Indicators (KPIs) for Environmental Criteria, which can help you measure and monitor your company's environmental performance:

- **Greenhouse gas (GHG) emissions:** Track your company's total GHG emissions, such as carbon dioxide (CO2), methane (CH4), and nitrous oxide (N2O), as well as emissions per unit of production or revenue. This KPI helps you monitor your progress in reducing your carbon footprint and combating climate change.

- **Energy consumption:** Measure the total energy consumed by your company, including electricity, natural gas, and other fuels, as well as energy usage per unit of output or revenue. This KPI can help you identify opportunities to improve energy efficiency and reduce energy costs.
- **Water usage:** Monitor the total volume of water consumed by your company, as well as water usage per unit of production or revenue. This KPI can help you assess your water efficiency and identify opportunities to conserve water resources.
- **Waste generation and management:** Track the amount of waste generated by your company, including solid waste, hazardous waste, and e-waste, as well as waste reduction, recycling, and diversion rates. This KPI can help you evaluate the effectiveness of your waste management practices and drive improvements in waste reduction and recycling efforts.
- **Renewable energy usage:** Measure the percentage of your company's energy consumption that comes from renewable sources, such as solar, wind, or hydroelectric power. This KPI can help you assess your progress in transitioning to cleaner energy sources and reducing your reliance on fossil fuels.
- **Biodiversity impact:** Monitor the impact of your company's operations on biodiversity, such as habitat preservation or restoration efforts, or the number of endangered species affected by your activities. This KPI can help you evaluate your efforts to protect and conserve ecosystems and wildlife.

Tracking these environmental KPIs will allow you to gain valuable insights into your company's environmental performance and identify areas where improvements can be made. This information can help you prioritize initiatives that align with your sustainability goals and demonstrate your commitment to responsible environmental practices.

Social Criteria

These criteria evaluate your company's relationships with its employees, customers, suppliers, and the communities in which it operates. Social KPIs may include:

- **Employee diversity:** Measure the representation of various demographic groups within your workforce, including gender, age, and ethnicity.
- **Health and safety:** Track workplace accident rates, lost workdays, and employee satisfaction with health and safety policies.
- **Community engagement:** Monitor the scope and impact of your company's philanthropic initiatives and partnerships with local organizations.

Here are some practical examples of Key Performance Indicators (KPIs) for Social Criteria, which can help you measure and monitor your company's social performance:

- **Employee diversity and inclusion:** Track the representation of various demographic groups within your workforce, such as gender, age, and ethnicity, as well as the implementation of diversity and inclusion initiatives. This KPI can help you assess the inclusiveness of your workplace and identify areas for improvement.
- **Employee satisfaction and retention:** Monitor employee satisfaction levels through regular surveys and feedback, as well as employee turnover rates. This KPI can help you evaluate the effectiveness of your human resource policies

and practices in fostering a positive work environment and retaining talent.

- **Health and safety:** Track workplace accident rates, lost workdays due to injuries or illnesses, and employee satisfaction with health and safety policies. This KPI can help you assess the effectiveness of your workplace safety programs and identify areas where improvements can be made.

- **Training and development:** Measure the number of training hours per employee, the percentage of employees participating in professional development programs, and the impact of training on employee performance. This KPI can help you evaluate the effectiveness of your employee development initiatives and identify opportunities for further growth and learning.

- **Supply chain responsibility:** Monitor your suppliers' adherence to social criteria, such as fair labor practices, human rights, and environmental responsibility, as well as your company's efforts to promote responsible sourcing. This KPI can help you assess the social and environmental impacts of your supply chain and identify areas for improvement.

- **Community engagement and philanthropy:** Track the scope and impact of your company's community engagement activities, such as volunteer hours, charitable donations, and partnerships with local organizations. This KPI can help you evaluate your company's contributions to the well-being of the communities in which it operates.

Tracking these social KPIs will allow you to gain valuable insights into your company's social performance and identify areas where improvements can be made. This information can help you prioritize initiatives that align with your social responsibility goals and demonstrate your commitment to ethical and responsible business practices.

Governance Criteria

These criteria examine your company's management structure, decision-making processes, and adherence to ethical standards. Governance KPIs may consist of:

- **Board diversity:** Assess the representation of different demographic groups and backgrounds among your company's board of directors.

- **Executive compensation:** Monitor the ratio of executive pay to median employee pay and evaluate the alignment of compensation structures with company performance.

- **Ethical conduct:** Track incidents of ethical violations, such as corruption or fraud, and assess the effectiveness of your company's compliance programs.

Identifying and tracking relevant ESG KPIs will allow you to gain valuable insights into your company's sustainability performance, enabling you to make data-driven decisions and prioritize initiatives that align with your ESG goals. This information can also help you communicate your progress to stakeholders, including investors, customers, and employees, demonstrating your commitment to responsible business practices.

Here are some practical examples of Key Performance Indicators (KPIs) for Governance Criteria, which can help you measure and monitor your company's governance performance:

- **Board diversity and independence:** Assess the representation of different demographic groups, backgrounds, and skill sets among your company's board of directors, as well as the proportion of independent board members. This KPI can help you evaluate the effectiveness of your board composition in promoting diverse perspectives and strong oversight.

- **Executive compensation:** Monitor the ratio of executive pay to median employee pay, the alignment of compensation structures with company performance, and the transparency of executive compensation policies. This KPI can help you evaluate the fairness of your compensation practices and ensure they align with shareholder interests.

- **Ethical conduct and compliance:** Track incidents of ethical violations, such as corruption, fraud, or conflicts of interest, and assess the effectiveness of your company's compliance programs, code of conduct, and internal controls. This KPI can help you evaluate your company's commitment to ethical business practices and identify areas for improvement.

- **Shareholder engagement:** Measure the level of communication and engagement between your company and its shareholders, including the frequency of shareholder meetings, investor presentations, and responsiveness to shareholder concerns. This KPI can help you assess the effectiveness of your investor relations efforts and ensure that shareholder interests are represented in decision-making processes.

- **Transparency and disclosure:** Monitor the quality, timeliness, and comprehensiveness of your company's financial and non-financial disclosures, such as annual reports, sustainability reports, and regulatory filings. This KPI can help you evaluate your company's commitment to transparency and its ability to provide stakeholders with the information they need to make informed decisions.

- **Risk management:** Assess your company's ability to identify, assess, and mitigate key risks, including financial, operational, strategic, and ESG-related risks. This KPI can help you evaluate the effectiveness of your risk management processes and ensure that your company is well-prepared to navigate potential challenges.

Tracking these governance KPIs will allow you to gain valuable insights into your company's governance performance and identify areas where improvements can be made. This information can help you prioritize initiatives that align with your governance goals and demonstrate your commitment to strong corporate governance practices.

Technology for Sustainability

Leveraging technology for sustainability is crucial to driving innovation, improving efficiency, and reducing your company's environmental impact. By harnessing the power of technology, you can identify and implement sustainable solutions across various aspects of your business operations. Here are some practical examples of how you can use technology to enhance your sustainability efforts:

- **Energy management systems:** Utilize advanced software solutions to monitor and analyze your company's energy consumption in real-time, identify inefficiencies, and optimize energy usage. These systems can help you reduce energy costs, improve efficiency, and lower your carbon footprint.

- **Internet of Things (IoT) devices:** Implement IoT devices, such as smart sensors and controllers, to automate and optimize various processes within your company. For example, you can use IoT devices to monitor and control lighting, heating, and cooling systems to maximize energy efficiency and minimize waste.

- **Big data and analytics:** Harness the power of big data and analytics to gain insights into your company's environmental, social, and governance (ESG) performance. By analyzing large datasets, you can identify patterns, trends, and areas for improvement, enabling you to make informed decisions about your sustainability initiatives.

- **Remote work technologies:** Promote remote work and virtual collaboration by adopting technologies such as video conferencing, cloud-based applications, and project management tools. Remote work can help you reduce your company's carbon footprint by decreasing the need for commuting and office space while also promoting work-life balance for your employees.

- **Circular economy technologies:** Explore technologies that support a circular economy, such as recycling, upcycling, and waste-to-energy conversion. These solutions can help you minimize waste, extend the lifespan of resources, and create new revenue streams from previously discarded materials.

- **Clean energy technologies:** Invest in renewable energy solutions, such as solar panels, wind turbines, or energy storage systems, to power your company's operations with clean, renewable energy sources. This can help you reduce your reliance on fossil fuels, lower your emissions, and contribute to a more sustainable energy future.

Incorporating these and other technology-driven solutions into your sustainability strategy will allow you to drive innovation, optimize your operations, and enhance your company's environmental performance. Embracing technology for sustainability can also help you differentiate your business from competitors, attract eco-conscious customers and investors, and create long-term value for your stakeholders.

Module 1 Exam Questions

1. What does ESG stand for?

a) Environmental, Social, Governance

b) Environmental, Systematic, Governance

c) Economic, Social, Governance

d) Environmental, Social, Globalization

2. Which of the following is a key aspect of Sustainability in business?

a) Value creation

b) Risk mitigation

c) Investor appeal

d) All of the above

3. Which KPI is commonly used to measure a company's environmental performance?

a) Greenhouse gas emissions

b) Employee satisfaction

c) Board diversity

d) Shareholder engagement

4. Which KPI is commonly used to measure a company's social performance?

a) Energy consumption

b) Employee diversity and inclusion

c) Executive compensation

d) Transparency and disclosure

5. Which KPI is commonly used to measure a company's governance performance?

a) Water usage

b) Health and safety

c) Board diversity and independence

d) Community engagement and philanthropy

6. What technology can be used to monitor and analyze a company's energy consumption in real-time?

a) Big data and analytics

b) Energy management systems

c) IoT devices

d) Remote work technologies

7. Which of the following is an example of a clean energy technology?

a) Solar panels

b) Wind turbines

c) Energy storage systems

d) All of the above

8. How can remote work technologies contribute to sustainability?

a) Reducing energy consumption

b) Decreasing the need for commuting

c) Minimizing office space requirements

d) All of the above

9. Which of the following is NOT a part of the social criteria in ESG?

a) Employee satisfaction

b) Supply chain responsibility

c) Greenhouse gas emissions

d) Community engagement

10. What is the main purpose of implementing ESG policies in a company?

a) To improve financial performance

b) To reduce environmental and social risks

c) To attract investors

d) To achieve regulatory compliance

11. In the context of ESG, what does "risk mitigation" refer to?

a) Minimizing financial risks

b) Reducing environmental and social risks

c) Avoiding legal risks

d) None of the above

12. What is the primary goal of sustainable investing?

a) To maximize financial returns

b) To minimize negative impacts on society

c) To make a positive contribution to society

d) All of the above

13. Which of the following is an example of a technology that supports a circular economy?

a) Recycling

b) Video conferencing

c) Energy management systems

d) IoT devices

14. Which KPI is commonly used to measure employee satisfaction and retention?

a) Employee turnover rate

b) Energy consumption

c) Risk management

d) Shareholder engagement

15. What is the primary benefit of using IoT devices in sustainability efforts?

a) Automating and optimizing processes

b) Reducing waste generation

c) Enhancing transparency

d) Improving communication

16. What is the main objective of impact investing?

a) To achieve high financial returns

b) To minimize risk

c) To contribute to solving global social problems

d) None of the above

17. What technology can help a company analyze large datasets to gain insights into its ESG performance?

a) Big data and analytics

b) Energy management systems

c) IoT devices

d) Remote work technologies

18. Which of the following is an example of a renewable energy source?

a) Solar power

b) Wind power

c) Hydroelectric power

d) All of the above

19. Which of the following factors can be positively influenced by incorporating ESG policies in a company?

a) Customer loyalty

b) Employee engagement

c) Competitive advantage

d) All of the above

20. What is a primary reason for measuring a company's ESG performance using Key Performance Indicators (KPIs)?

a) To improve internal communication

b) To identify areas for improvement

c) To increase brand awareness

d) To reduce operational costs

Answers:

1. a
2. d
3. a
4. b
5. c
6. b
7. d
8. d
9. c
10. b
11. b
12. c
13. a
14. a
15. a
16. c
17. a
18. d
19. d
20. b

Module 2. Environmental Sustainability

Introduction to Environmental Sustainability

Welcome to Module 2 on Environmental Sustainability, where we will explore the critical role of environmental management and conservation in building a sustainable future. In this module, we will delve into the key principles of environmental sustainability and offer practical examples to illustrate how businesses can adopt eco-friendly practices that promote long-term prosperity while minimizing their impact on the planet.

Environmental sustainability is an essential pillar of the ESG (Environmental, Social, and Governance) framework, which seeks to create value for businesses, society, and the environment. It involves taking a responsible and proactive approach to addressing environmental challenges, such as climate change, resource depletion, pollution, and habitat destruction. By adopting sustainable practices, businesses can not only reduce their environmental footprint but also unlock new opportunities, enhance their reputation, and drive innovation.

One practical example of environmental sustainability in action is the transition to renewable energy sources, such as solar, wind, and hydroelectric power. By investing in clean energy technologies, businesses can reduce their dependence on fossil fuels and lower greenhouse gas emissions, which contribute to climate change. This approach not only benefits the environment but also helps companies to secure long-term energy supplies and reduce operational costs.

Another example is the adoption of circular economy principles, where businesses strive to minimize waste by reusing, recycling, and repurposing materials and products throughout their life cycles. This approach promotes resource efficiency, reduces the demand for raw materials, and helps businesses tap into new markets and revenue streams.

Furthermore, businesses can also protect and restore ecosystems by implementing sustainable land-use practices and investing in habitat conservation projects. For instance, companies can establish partnerships with local communities and environmental organizations to preserve biodiversity and create a shared vision for a sustainable future.

In this module, we will dive deeper into the various aspects of environmental sustainability and explore best practices and strategies that businesses can employ to address key challenges and capitalize on emerging opportunities. Through a combination of theoretical knowledge and real-world examples, we aim to equip you with the tools and insights needed to become a leader in environmental sustainability and drive positive change in your organization and beyond.

Stay tuned as we embark on a journey to discover the importance of environmental sustainability and how it can contribute to the long-term success and resilience of businesses in an increasingly interconnected and resource-constrained world.

Climate Change and Greenhouse Gas Emissions

Climate change is one of the most pressing global challenges of our time, with far-reaching consequences for ecosystems, economies, and communities. As a result, addressing climate change and reducing greenhouse gas (GHG) emissions are crucial aspects of environmental sustainability. In this section, we will explore the relationship between climate change and GHG emissions, as well as the strategies you can employ to minimize your organization's impact on the environment.

Greenhouse gases, such as carbon dioxide (CO_2), methane (CH_4), and nitrous oxide (N_2O), trap heat in the Earth's atmosphere, leading to a rise in global temperatures. The majority of GHG emissions stem from human activities, including burning fossil fuels for energy production, deforestation, and industrial processes. By reducing your organization's GHG emissions, you can help mitigate climate change and contribute to a more sustainable future.

To reduce your organization's GHG emissions, consider implementing the following strategies:

- **Energy efficiency:** Increasing the energy efficiency of your operations, buildings, and equipment can significantly reduce your GHG emissions. Simple actions such as upgrading lighting systems, optimizing heating, ventilation, and air conditioning (HVAC) systems, and investing in energy-efficient appliances can result in substantial energy savings and emission reductions.

- **Renewable energy:** Transitioning to renewable energy sources, such as solar, wind, or hydroelectric power, can help you lower your organization's carbon footprint. You can either install renewable energy systems on-site or purchase renewable energy certificates (RECs) to support clean energy production.

- **Transportation:** Encourage sustainable transportation options for your employees, such as carpooling, biking, or using public transit. Additionally, consider investing in electric or hybrid vehicles for your company fleet and providing charging infrastructure to support their use.

- **Supply chain management:** Engage with your suppliers to promote sustainable practices and reduce the carbon footprint of your supply chain. This could include sourcing raw materials from certified sustainable sources, optimizing logistics to reduce transportation emissions, or partnering with suppliers that prioritize energy efficiency and renewable energy.

- **Carbon offsetting:** While reducing emissions should be your primary focus, you can also offset any remaining emissions by investing in certified carbon offset projects that reduce GHG emissions elsewhere. This could include supporting reforestation efforts, renewable energy projects, or methane capture initiatives.

Incorporating these strategies into your organization's operations will allow you to take significant steps toward reducing your greenhouse gas emissions and mitigating the effects of climate change. Additionally, you will be better positioned to meet regulatory requirements, attract environmentally conscious customers and investors, and strengthen your organization's reputation as a leader in sustainability.

Practical examples of energy efficiency

Here are some practical examples of how you can improve energy efficiency in your operations, buildings, and equipment to reduce GHG emissions:

- **Lighting upgrades:** Replace traditional incandescent or halogen bulbs with energy-efficient LED lighting, which can consume up to 90% less energy and last up to 25 times longer. You can also install occupancy sensors and daylight harvesting systems to ensure that lights are only used when needed, further reducing energy consumption.

- **HVAC optimization:** Regular maintenance and proper management of your heating, ventilation, and air conditioning (HVAC) systems can significantly improve energy efficiency. This may include cleaning or replacing air filters, sealing duct leaks, and installing programmable thermostats to optimize temperature settings based on occupancy and time of day. Additionally, consider upgrading to high-efficiency HVAC equipment that uses less energy and emits fewer GHGs.

- **Building envelope improvements:** Enhance your building's insulation, windows, and doors to minimize heat loss in the winter and heat gain in the summer. This can be achieved through measures such as adding insulation to walls and attics, installing double-glazed or triple-glazed windows, and using weatherstripping to seal gaps around doors and windows.

- **Energy-efficient appliances:** When purchasing new equipment or appliances for your organization, look for those with ENERGY STAR certification or other recognized efficiency ratings. These products are designed to use less energy without sacrificing performance, helping you save on energy costs and reduce GHG emissions.

- **Energy management systems:** Implement a building energy management system (BEMS) to monitor, control, and optimize your facility's energy consumption. This technology can help you identify inefficiencies, set energy-saving targets, and track progress over time.

- **Employee engagement:** Encourage your employees to adopt energy-saving habits, such as turning off lights and electronic devices when not in use, using natural lighting when possible, and adjusting thermostats to save energy during non-working hours.

Implementing these energy efficiency measures will allow you to significantly reduce your organization's GHG emissions and energy costs, contributing to a more sustainable future and demonstrating your commitment to environmental stewardship.

Practical examples of renewable energy

Here are some practical examples of how your organization can transition to renewable energy sources and lower its carbon footprint:

- **On-site solar power:** Install solar photovoltaic (PV) panels on your building's rooftop or on available land to generate clean electricity for your organization's use. This can help you reduce reliance on grid electricity and lower GHG emissions. Additionally, any excess energy generated can be fed back into the grid, providing potential income through net metering or feed-in tariff schemes.

- **Wind energy:** If your organization's location has access to adequate wind resources, consider installing wind turbines to harness wind power for your energy needs. Small-scale wind turbines can be used to supplement your electricity supply or even power remote facilities that are not connected to the grid.

- **Solar thermal systems:** Implement solar thermal systems to heat water for your organization's use, reducing the need for conventional water heating methods that rely on fossil fuels. Solar thermal collectors can be installed on rooftops or walls and can be used to support heating, ventilation, and air conditioning (HVAC) systems or industrial processes that require hot water.

- **Renewable energy procurement:** If on-site renewable energy generation is not feasible, consider purchasing renewable energy certificates (RECs) or entering into power purchase agreements (PPAs) with renewable energy providers. These options allow you to support clean energy production without having to install and maintain your own renewable energy systems.

- **Green tariffs:** Some utility companies offer green tariffs or renewable energy plans that allow you to source a portion or all of your electricity from renewable sources. By choosing a green tariff, you can directly support the growth of clean energy in your region and reduce your organization's carbon footprint.

- **Energy storage:** To maximize the benefits of renewable energy systems, consider integrating energy storage solutions, such as batteries or thermal storage systems. Energy storage can help you store excess renewable energy generated during periods of high production and use it during periods of low production or peak demand, further reducing your reliance on grid electricity.

Adopting these renewable energy strategies will allow you to significantly lower your organization's carbon footprint, demonstrate your commitment to sustainability, and potentially save on energy costs in the long run. Additionally, transitioning to renewable energy sources can help you meet regulatory requirements and appeal to environmentally conscious customers and investors.

Practical examples of transportation

Here are some practical examples of how you can promote sustainable transportation options for your employees and invest in greener vehicles for your company fleet:

- **Carpooling program:** Establish a carpooling program within your organization to help employees connect and share rides to work. This can reduce the number of single-occupancy vehicles on the road, leading to lower GHG emissions and less traffic congestion. You can facilitate carpooling through online platforms, ride-sharing apps, or designated carpool parking spaces.

- **Bike-friendly workplace:** Encourage employees to cycle to work by providing secure bike storage, showers, and changing facilities. You can also offer incentives, such as subsidized bike purchases, maintenance workshops, or participation in bike-to-work programs to further promote cycling as a healthy and eco-friendly transportation option.
- **Public transit incentives:** Offer subsidies or pre-tax benefits to employees who use public transportation to commute to work. This can encourage the use of buses, trains, and other mass transit systems, reducing the number of single-occupancy vehicles on the road and lowering overall GHG emissions.
- **Electric and hybrid company vehicles:** Replace conventional gasoline or diesel vehicles in your company fleet with electric or hybrid models to reduce emissions and fuel consumption. This not only lowers your organization's carbon footprint but also demonstrates your commitment to sustainable transportation.
- **Charging infrastructure:** Install electric vehicle (EV) charging stations at your workplace to support employees who own or use electric vehicles. By providing convenient and accessible charging options, you can encourage the adoption of EVs and contribute to the growth of the electric vehicle market.
- **Telecommuting and flexible work arrangements:** Implement telecommuting policies and flexible work arrangements that allow employees to work remotely, reducing the need for daily commutes. By cutting down on

the number of trips employees make to the office, you can help reduce overall transportation-related GHG emissions.

- **Eco-driving training:** Provide training and resources to help your employees adopt eco-driving habits, such as maintaining consistent speeds, avoiding rapid acceleration and braking, and properly inflating tires. These practices can lead to fuel savings and reduced emissions, regardless of the type of vehicle being used.

Promoting sustainable transportation options and investing in greener company vehicles will allow you to significantly reduce your organization's transportation-related GHG emissions, save on fuel costs, and demonstrate your commitment to environmental sustainability.

Practical examples of supply chain management

Here are some practical examples of how you can engage with your suppliers to promote sustainable practices and reduce the carbon footprint of your supply chain:

- **Sustainable sourcing:** Source raw materials and products from suppliers that adhere to certified sustainable standards, such as the Forest Stewardship Council (FSC) for wood and paper products, Fair Trade for agricultural goods, or the Responsible Business Alliance (RBA) for electronics. This can help ensure that your suppliers follow environmentally friendly and socially responsible practices.

- **Supplier assessments:** Conduct regular sustainability assessments of your suppliers to evaluate their environmental performance and identify areas for improvement. Encourage suppliers to adopt sustainable practices, such as reducing waste, improving energy efficiency, and using renewable energy.

- **Local sourcing:** Whenever possible, source materials and products from local suppliers to reduce transportation emissions and support local economies. This can help minimize the environmental impact of your supply chain while also fostering stronger relationships with your suppliers.

- **Green logistics:** Collaborate with your suppliers and logistics partners to optimize transportation routes, consolidate shipments, and use more fuel-efficient vehicles or modes of transport. These measures can help reduce transportation-related emissions and improve overall supply chain efficiency.
- **Supplier training and capacity building:** Provide training and resources to help your suppliers adopt sustainable practices and technologies. This can include sharing best practices, offering technical assistance, or hosting workshops on topics such as energy efficiency, waste reduction, or sustainable sourcing.
- **Supplier recognition and incentives:** Recognize and reward suppliers that demonstrate strong environmental performance and commitment to sustainability. This can include offering financial incentives, public recognition, or preferential treatment in contract negotiations to encourage ongoing improvement.
- **Collaborative partnerships:** Establish long-term partnerships with suppliers that share your commitment to sustainability. Work together to set joint sustainability goals, share best practices, and leverage each other's expertise to drive continuous improvement and innovation in your supply chain.

Engaging with your suppliers and promoting sustainable practices throughout your supply chain will allow you to significantly reduce your organization's environmental impact, demonstrate your commitment to sustainability, and potentially unlock new business opportunities. This proactive approach can also help you manage risks, enhance your reputation, and appeal to environmentally conscious customers and investors.

Practical examples of carbon offsetting

Here are some practical examples of how you can invest in certified carbon offset projects to compensate for your organization's remaining emissions:

- **Reforestation and afforestation projects:** Support tree planting initiatives or projects that restore or establish new forests, which help absorb carbon dioxide from the atmosphere. By investing in these projects, you can offset your organization's emissions while also contributing to biodiversity conservation and ecosystem restoration.
- **Renewable energy projects:** Invest in projects that generate clean energy from sources like solar, wind, or hydropower. By supporting the development and expansion of renewable energy infrastructure, you can offset your organization's emissions and contribute to the transition away from fossil fuels.
- **Methane capture and utilization:** Support projects that capture and utilize methane from landfills, agricultural waste, or other sources. Methane is a potent greenhouse gas with a much higher global warming potential than carbon dioxide. By capturing and using this methane for energy production or other purposes, you can offset your organization's emissions and help reduce overall GHG levels.
- **Energy efficiency projects:** Invest in projects that improve energy efficiency in homes, businesses, or industrial facilities. These projects can help reduce the overall demand for energy and the associated GHG emissions, contributing to your organization's offset efforts.

- **Clean cookstove initiatives:** Support projects that provide clean cookstoves to communities in developing countries, replacing traditional open-fire cooking methods that release large amounts of carbon dioxide and other harmful pollutants. By investing in clean cookstove initiatives, you can offset your organization's emissions while improving the health and quality of life for people in these communities.

- **Carbon capture and storage (CCS) projects:** Invest in projects that capture and store carbon dioxide emissions from industrial processes or power generation. CCS technology can help prevent large amounts of carbon dioxide from being released into the atmosphere, offsetting your organization's emissions.

- **Sustainable agriculture projects:** Support projects that promote sustainable agriculture practices, such as agroforestry, organic farming, or regenerative agriculture. These practices can help sequester carbon in soil and vegetation, offsetting your organization's emissions and promoting long-term environmental and social benefits.

When selecting carbon offset projects, it is essential to ensure they are certified by reputable standards, such as the Verified Carbon Standard (VCS), Gold Standard, or the Climate Action Reserve. This helps ensure the projects' credibility and the effectiveness of your investment in reducing GHG emissions. While carbon offsetting should not replace your organization's efforts to reduce emissions directly, it can be a valuable tool for compensating for unavoidable emissions and demonstrating your commitment to environmental sustainability.

Practical KPIs for Climate Change and Greenhouse Gas Emissions

The following KPIs can be used for tracking/managing progress towards reducing greenhouse gas emissions and addressing climate change:

1. **Total greenhouse gas emissions:** Measure the total amount of greenhouse gas emissions produced by your organization, typically expressed in metric tons of carbon dioxide equivalent (CO2e) per year.

2. **Greenhouse gas emissions intensity:** Calculate the greenhouse gas emissions intensity by dividing the total emissions by a relevant factor, such as revenue, square footage, or production output. This helps to normalize the data and allows for comparison across different organizations or industries.

3. **Energy consumption:** Monitor the total amount of energy consumed by your organization, typically measured in kilowatt-hours (kWh) or megawatt-hours (MWh) for electricity and joules or British thermal units (BTUs) for other energy sources.

4. **Energy efficiency:** Measure the energy efficiency of your operations by calculating the ratio of output (products, services, or value created) to the energy input required.

5. **Renewable energy usage:** Track the percentage of your organization's total energy consumption that comes from renewable sources, such as solar, wind, or hydroelectric power.

6. **Carbon footprint reduction:** Monitor the progress in reducing your organization's overall carbon footprint over time, typically measured as a percentage decrease from a baseline year.

7. **Scope 1, 2, and 3 emissions:** Measure and report on direct emissions (Scope 1), indirect emissions from purchased electricity (Scope 2), and other indirect emissions from sources such as supply chain, employee commuting, or business travel (Scope 3).

8. **Carbon offsetting:** Track the amount of certified carbon offset credits your organization has purchased or generated through supporting projects that reduce greenhouse gas emissions elsewhere.

Resource Management and Waste Reduction

Effective resource management and waste reduction are essential components of environmental sustainability. By using resources more efficiently and minimizing waste, you can reduce your environmental impact, save money, and enhance your reputation for sustainability. Here are some strategies and practical examples to help you manage resources and reduce waste:

- **Reduce, Reuse, Recycle:** Adopt the "3 Rs" approach to manage waste in your operations. First, focus on reducing the amount of waste generated by optimizing processes, eliminating unnecessary packaging, and using digital alternatives to paper. Second, reuse materials and equipment whenever possible, extending their lifespan and reducing the need for new resources. Finally, implement recycling programs to ensure waste materials are repurposed or recycled.

  Practical example: Implement a company-wide recycling program by placing clearly labeled recycling bins throughout your facilities and providing training on proper recycling practices.

- **Circular economy:** Transition from a linear "take-make-waste" model to a circular economy approach, where resources are used, recovered, and regenerated at the end of their service life. Design products and processes with the goal of minimizing waste, maximizing resource efficiency, and enabling easy disassembly and recycling.

  Practical example: Work with your product design team to create modular products that can be easily repaired, upgraded, or disassembled for recycling at the end of their life.

- **Waste audits:** Conduct regular waste audits to identify the types and amounts of waste generated in your operations. Use this data to develop targeted waste reduction strategies and track your progress over time.

  Practical example: Perform a waste audit by measuring and categorizing the waste generated in your facilities over a set period, then set targets for waste reduction based on your findings.

- **Sustainable procurement:** Incorporate sustainability criteria into your procurement processes, prioritizing products and materials with lower environmental impacts, such as recycled content, biodegradable materials, or products with extended lifespans.

  Practical example: Establish a green purchasing policy that outlines sustainability criteria for products and materials, and train your procurement team to apply these criteria when selecting suppliers.

- **Resource-efficient technologies:** Invest in resource-efficient technologies and equipment that can help you reduce waste and save resources, such as energy-efficient appliances, water-saving fixtures, and waste reduction systems.

  Practical example: Upgrade to energy-efficient lighting systems, such as LED lights, and install motion sensors to ensure lights are only used when needed.

- **Employee engagement:** Involve your employees in your resource management and waste reduction efforts, providing training and resources to help them adopt sustainable practices at work.

  Practical example: Organize a waste reduction challenge, where employees compete to reduce waste in their departments, with rewards for the most successful teams.

Implementing these resource management and waste reduction strategies will allow you to significantly reduce your environmental impact, improve operational efficiency, and demonstrate your commitment to sustainability.

Practical KPIs for Resource Management and Waste Reduction

The following KPIs can be used to track/measure progress in resource management and waste reduction efforts, helping to identify areas for improvement and guide decision-making:

1. **Total waste generated:** Measure the total amount of waste produced by your organization, typically expressed in metric tons or kilograms.

2. **Waste diversion rate:** Calculate the percentage of waste diverted from landfills or incineration through recycling, composting, or other waste reduction efforts.

3. **Recycling rate:** Monitor the percentage of waste materials that are collected and recycled within your organization.

4. **Water consumption:** Measure the total amount of water consumed by your organization, typically expressed in cubic meters or liters.

5. **Water efficiency:** Calculate the ratio of water used to the output produced (such as products, services, or value created) to assess your organization's water efficiency.

6. **Material usage efficiency:** Monitor the efficiency of material usage in your organization's production processes, such as the percentage of raw materials converted into finished products.

7. **Packaging reduction:** Track the progress in reducing packaging materials used in your organization, typically measured as a percentage decrease from a baseline year.

8. **Paper usage:** Measure the total amount of paper consumed by your organization, and track the percentage of recycled or sustainably sourced paper used.

9. **Hazardous waste reduction:** Monitor the amount of hazardous waste generated by your organization, and track the progress in reducing its production over time.

10. **Energy consumption:** Measure the total energy consumed by your organization, typically in kilowatt-hours (kWh) or megawatt-hours (MWh) for electricity, and joules or British thermal units (BTUs) for other energy sources. This KPI can be used in conjunction with waste reduction efforts, as improved resource management often leads to reduced energy consumption.

Biodiversity and and Ecosystem Preservation

Biodiversity and ecosystem preservation are critical aspects of environmental sustainability. Protecting and enhancing biodiversity ensures the health and resilience of ecosystems, which provide essential services like clean air and water, climate regulation, and soil fertility. As a responsible business, you can contribute to biodiversity conservation and ecosystem preservation through the following strategies and practical examples:

- **Habitat protection:** Identify and protect any natural habitats on or near your premises, taking steps to minimize your impact on these ecosystems. You can also support habitat conservation projects or initiatives in your local area or within your supply chain.

  Practical example: Implement a no-deforestation policy and ensure that any new construction projects on your premises avoid sensitive habitats or minimize their impact on the surrounding environment.

- **Sustainable land use:** Adopt sustainable land use practices that promote biodiversity conservation, soil health, and water quality. This could include implementing reforestation projects, preserving wetlands, or practicing regenerative agriculture.

  Practical example: If your business operates in the agricultural sector, use crop rotation, cover crops, and no-till farming techniques to promote soil health and reduce erosion, while supporting local biodiversity.

- **Invasive species management:** Monitor and control invasive species on your premises to prevent their spread and protect native biodiversity. Collaborate with local authorities or environmental organizations to share information and resources related to invasive species management.

Practical example: Establish a monitoring program to identify and remove invasive plant species from your property, and replant with native species that support local biodiversity.

- **Supply chain management:** Engage with your suppliers to promote biodiversity-friendly practices throughout your supply chain. Prioritize suppliers that demonstrate a commitment to biodiversity conservation and sustainable land use.

  Practical example: Require suppliers of agricultural products to adhere to sustainable farming practices that support biodiversity, such as integrated pest management, organic farming, or agroforestry.

- **Ecosystem services valuation:** Assess the value of the ecosystem services provided by the natural habitats on or near your premises, and incorporate this value into your decision-making processes. This can help you make more informed decisions about land use, resource management, and conservation efforts.

  Practical example: Conduct an ecosystem services assessment to determine the economic value of services like water filtration, carbon sequestration, and pollination provided by the ecosystems surrounding your facilities.

- **Employee engagement and education:** Encourage your employees to participate in biodiversity conservation and ecosystem preservation efforts by providing training, resources, and opportunities to get involved.

  Practical example: Organize a tree-planting event for employees to participate in, fostering a sense of responsibility for the environment while contributing to local reforestation efforts.

Integrating biodiversity and ecosystem preservation into your business practices will allow you to contribute to the health and resilience of our planet, while also demonstrating your commitment to environmental sustainability. This can enhance your reputation and appeal to customers, investors, and other stakeholders who value responsible business practices.

Practical KPIs for Biodiversity and Ecosystem Preservation

The following KPIs can be used to track/measure progress in biodiversity and ecosystem preservation efforts, providing insights into the health and resilience of ecosystems and guiding actions to improve their conservation:

1. **Habitat conservation:** Track the total area of land (in hectares or acres) that your organization has conserved or restored to support biodiversity and ecosystem preservation.

2. **Native species protection:** Monitor the number of native species protected or reintroduced through your organization's biodiversity initiatives.

3. **Ecosystem services valuation:** Calculate the economic value of ecosystem services (such as carbon sequestration, water filtration, and pollination) provided by the habitats conserved or restored by your organization.

4. **Sustainable land use:** Measure the percentage of your organization's land that is managed in accordance with sustainable land-use practices, such as organic farming, agroforestry, or wildlife-friendly landscaping.

5. **Invasive species management:** Track the progress in controlling invasive species on your organization's land or

within its supply chain, typically measured as a reduction in the number of invasive species or their impact on native ecosystems.

6. **Biodiversity risk assessment:** Conduct regular assessments to identify and prioritize the biodiversity risks associated with your organization's operations, supply chain, and products.

7. **Supply chain biodiversity:** Monitor the percentage of suppliers that implement biodiversity and ecosystem preservation measures, such as certified sustainable sourcing or habitat conservation initiatives.

8. **Employee engagement in biodiversity initiatives:** Track the number or percentage of employees who participate in biodiversity-related volunteer programs, training, or awareness campaigns.

9. **Partnerships for biodiversity conservation:** Measure the number of partnerships or collaborations your organization has established with NGOs, academic institutions, or other stakeholders to support biodiversity and ecosystem preservation.

10. **Biodiversity-related certifications or standards:** Monitor the progress in achieving and maintaining biodiversity-related certifications or standards, such as Forest Stewardship Council (FSC) certification, Rainforest Alliance certification, or adherence to the Convention on Biological Diversity (CBD) guidelines.

Pollution Prevention and Control

Preventing and controlling pollution is essential for environmental sustainability and the health of our planet. Pollution can negatively impact air, water, and soil quality, as well as the well-being of humans and other living organisms. As a responsible business, you can adopt various strategies and practical examples to prevent and control pollution in your operations:

- **Emissions reduction:** Identify and monitor sources of air pollution within your operations, such as vehicle fleets, manufacturing processes, or energy consumption. Implement strategies to reduce emissions, including investing in energy-efficient technologies, adopting cleaner production processes, or using low-emission fuels.

  Practical example: Upgrade your company fleet to electric or hybrid vehicles, reducing tailpipe emissions and improving local air quality.

- **Water pollution prevention:** Implement measures to prevent water pollution from your operations, such as proper waste disposal, wastewater treatment, and stormwater management. Regularly monitor and test the quality of water discharged from your facilities to ensure compliance with environmental regulations.

  Practical example: Install a wastewater treatment system at your facility to treat and remove pollutants before discharging the water, ensuring it meets regulatory standards and does not harm local water bodies.

- **Waste management:** Properly dispose of hazardous and non-hazardous waste generated by your operations, following local regulations and best practices. Consider recycling or repurposing waste materials whenever possible to minimize pollution.

Practical example: Set up a hazardous waste management program to ensure the safe collection, storage, and disposal of chemicals, solvents, or other hazardous materials used in your operations.

- **Pollution prevention planning:** Develop a comprehensive pollution prevention plan for your business, outlining strategies and targets for reducing pollution across your operations. Regularly review and update the plan to reflect changes in your business activities, technologies, and regulations.

  Practical example: Create a pollution prevention plan that sets targets for reducing air emissions, water pollution, and hazardous waste generation, and outlines the specific actions and investments needed to achieve these targets.

- **Employee training and engagement:** Provide training and resources to help employees understand their role in preventing pollution and the steps they can take to minimize their environmental impact at work.

  Practical example: Offer training workshops on pollution prevention, waste management, and environmental best practices to ensure employees understand their responsibilities and have the knowledge and tools to contribute to pollution reduction efforts.

- **Supply chain management:** Engage with your suppliers to promote pollution prevention and control throughout your supply chain. Prioritize suppliers that demonstrate a commitment to reducing pollution and complying with environmental regulations.

  Practical example: Develop a supplier code of conduct that includes pollution prevention requirements, and conduct regular audits to ensure suppliers adhere to these standards.

Incorporating pollution prevention and control measures into your business operations will allow you to minimize your environmental impact, reduce costs associated with waste disposal and regulatory compliance, and demonstrate your commitment to environmental sustainability. This can enhance your reputation and attract customers, investors, and other stakeholders who value responsible business practices.

Practical KPIs for Pollution Prevention and Control

The following KPIs can be used to track/measure progress in pollution prevention and control efforts, helping to identify sources of pollution, assess the effectiveness of control measures, and guide decision-making to improve environmental quality:

1. **Air pollutant emissions:** Measure the total amount of air pollutant emissions generated by your organization, such as nitrogen oxides (NOx), sulfur oxides (SOx), and particulate matter (PM).

2. **Emissions intensity:** Calculate the emissions intensity by dividing the total air pollutant emissions by a relevant factor, such as production output or revenue, to allow for comparisons across organizations or industries.

3. **Water pollution reduction:** Monitor the amount of pollutants discharged into water bodies by your organization, such as chemical oxygen demand (COD), biological oxygen demand (BOD), or total suspended solids (TSS).

4. **Wastewater treatment efficiency:** Measure the efficiency of your organization's wastewater treatment processes, typically expressed as a percentage of pollutants removed from the wastewater before discharge.

5. **Noise pollution reduction:** Monitor noise levels generated by your organization's operations and track progress in reducing noise pollution over time.

6. **Hazardous materials management:** Measure the total amount of hazardous materials used or generated by your organization and monitor progress in reducing their use, proper disposal, or substitution with less harmful alternatives.

7. **Spill prevention and response:** Track the number of spills or releases of hazardous materials, chemicals, or pollutants, and assess the effectiveness of your organization's spill prevention and response measures.

8. **Environmental permits and compliance:** Monitor the number of environmental permits held by your organization and track any instances of non-compliance or regulatory enforcement actions.

9. **Pollution prevention initiatives:** Track the number and effectiveness of pollution prevention initiatives implemented by your organization, such as process modifications, technology upgrades, or employee training programs.

10. **Air and water quality monitoring:** Regularly monitor and report on air and water quality at your organization's facilities and surrounding areas to ensure compliance with environmental standards and identify potential pollution issues.

Sustainable Procurement and Supply Chain Management

Sustainable procurement and supply chain management involve considering environmental, social, and economic factors when selecting suppliers and purchasing goods and services. By adopting sustainable procurement practices, you can reduce the environmental impact of your supply chain, support ethical business practices, and create long-term value for your organization. Here are some strategies and practical examples to help you implement sustainable procurement and supply chain management:

- **Supplier assessment:** Evaluate your suppliers based on their environmental, social, and governance (ESG) performance, prioritizing those that demonstrate a commitment to sustainability, ethical business practices, and transparent reporting.

  Practical example: Develop a supplier scorecard that rates suppliers on their ESG performance, including factors like energy efficiency, waste reduction, fair labor practices, and anti-corruption policies.

- **Sustainable sourcing:** Prioritize the procurement of goods and services that have a lower environmental impact, such as products made from recycled materials, certified sustainable timber, or organic food ingredients.

  Practical example: Update your procurement policy to require the purchase of office supplies made from recycled materials, such as paper, pens, and furniture.

- **Green contracts:** Incorporate sustainability clauses into your contracts with suppliers, setting clear expectations for environmental performance, compliance with regulations, and continuous improvement.

Practical example: Include clauses in your contracts requiring suppliers to implement waste reduction and recycling programs, adhere to energy efficiency standards, and report on their environmental performance.

- **Collaboration and capacity building:** Work closely with your suppliers to support their sustainability efforts, providing training, resources, and incentives to help them improve their environmental performance and adopt sustainable practices.

  Practical example: Organize a supplier workshop focused on energy efficiency and waste reduction, sharing best practices and providing technical assistance to help suppliers implement improvements.

- **Monitoring and reporting:** Regularly monitor and evaluate the sustainability performance of your suppliers, using tools such as supplier audits, surveys, and self-assessments. Communicate your supply chain sustainability achievements to stakeholders through transparent reporting.

  Practical example: Conduct annual supplier audits to verify compliance with your sustainability requirements, and include a summary of your supply chain sustainability performance in your annual sustainability report.

- **Risk management:** Assess the environmental and social risks associated with your supply chain, such as resource scarcity, climate change impacts, or labor rights violations. Develop strategies to mitigate these risks, such as diversifying your supplier base, investing in more sustainable materials, or collaborating with industry peers to address shared challenges.

  Practical example: Conduct a supply chain risk assessment to identify potential vulnerabilities related to water scarcity, and develop a water stewardship strategy to ensure the sustainable use of water resources throughout your supply chain.

Implementing sustainable procurement and supply chain management practices will allow you to reduce your environmental impact, support responsible business practices, and create long-term value for your organization. This can enhance your reputation, attract customers and investors who value sustainability, and contribute to a more sustainable global economy.

Practical KPIs for Sustainable Procurement and Supply Chain Management

The following KPIs can be used to track/measure progress in sustainable procurement and supply chain management, helping to identify areas for improvement, monitor supplier performance, and guide decision-making towards more sustainable practices:

1. **Sustainable sourcing:** Measure the percentage of materials, products, or services sourced from suppliers that meet your organization's sustainability criteria, such as environmental certifications or fair labor practices.

2. **Supplier sustainability assessments:** Track the number or percentage of suppliers assessed for sustainability performance, and monitor the improvements made by suppliers over time.

3. **Supply chain carbon footprint:** Calculate the total greenhouse gas emissions associated with your organization's supply chain, typically expressed in metric tons of CO2 equivalent (tCO2e).

4. **Ethical labor practices:** Measure the percentage of suppliers that comply with ethical labor standards, such as fair wages, safe working conditions, and no child or forced labor.

5. **Circular economy initiatives:** Track the implementation of circular economy principles in your supply chain, such as product take-back programs, recycling or upcycling initiatives, or designing products for durability and easy maintenance.

6. **Local sourcing:** Monitor the percentage of materials, products, or services sourced locally to reduce transportation emissions and support local economies.

7. **Supplier diversity:** Measure the percentage of your organization's procurement spending with diverse suppliers, such as minority-owned, women-owned, or small businesses.

8. **Conflict-free materials:** Track the use of conflict-free materials, such as minerals sourced from conflict-free zones, in your organization's products.

9. **Sustainable packaging:** Monitor the percentage of packaging materials that are recyclable, reusable, or made from renewable resources in your organization's supply chain.

10. **Supply chain transparency:** Assess the level of transparency in your organization's supply chain, such as the ability to trace products back to their source or verify the sustainability claims made by suppliers.

Environmental Management Systems

An Environmental Management System (EMS) is a structured approach to managing your organization's environmental performance. It involves setting environmental objectives, implementing policies and procedures, monitoring performance, and continuously improving your processes to minimize your environmental impact. By implementing an EMS, you can enhance your environmental performance, reduce costs, ensure compliance with regulations, and demonstrate your commitment to sustainability. Here are some strategies and practical examples to help you develop and implement an EMS:

- **EMS framework:** Adopt a recognized EMS framework, such as ISO 14001 or the Eco-Management and Audit Scheme (EMAS), which provides guidelines and best practices for managing your environmental performance.

  Practical example: Obtain ISO 14001 certification for your organization, demonstrating your commitment to environmental management and continuous improvement.

- **Environmental policy:** Develop a clear and comprehensive environmental policy that outlines your commitment to minimizing your environmental impact, complying with regulations, and continuously improving your performance.

  Practical example: Draft an environmental policy that sets targets for reducing greenhouse gas emissions, improving energy efficiency, and minimizing waste generation.

- **Objectives and targets:** Set specific, measurable, achievable, relevant, and time-bound (SMART) environmental objectives and targets for your organization, aligned with your overall business strategy and environmental policy.

Practical example: Establish a target to reduce your energy consumption by 20% within the next three years by implementing energy efficiency measures and investing in renewable energy.

- **Roles and responsibilities:** Assign clear roles and responsibilities for environmental management within your organization, ensuring that all employees understand their obligations and have the necessary resources and training to fulfill them.

  Practical example: Appoint an Environmental Management Representative (EMR) to oversee the implementation of your EMS, coordinate training, and monitor performance.

- **Monitoring and measurement:** Implement systems for monitoring and measuring your environmental performance, using key performance indicators (KPIs) to track progress towards your objectives and targets.

  Practical example: Monitor your energy consumption, waste generation, and water usage on a monthly basis, using this data to identify trends, evaluate progress towards your targets, and identify opportunities for improvement.

- **Legal compliance:** Establish processes for identifying, tracking, and complying with all applicable environmental laws and regulations, as well as any voluntary commitments or standards your organization adheres to.

  Practical example: Create a legal register that lists all relevant environmental laws and regulations, and conduct regular compliance audits to ensure your organization is meeting its legal obligations.

- **Continuous improvement:** Regularly review and update your EMS to identify opportunities for improvement, address gaps in performance, and respond to changes in your business environment, technologies, or regulations.

  Practical example: Conduct an annual management review of your EMS, evaluating your performance against your objectives and targets, and updating your EMS as needed to address new opportunities or challenges.

Implementing an EMS will allow you to effectively manage your environmental performance, reduce your environmental impact, and demonstrate your commitment to sustainability. This can enhance your reputation, attract customers and investors who value responsible business practices, and contribute to a more sustainable future.

Practical KPIs for Environmental Management Systems

The following KPIs can be used to track/measure progress in implementing and improving environmental management systems, providing insights into environmental performance, identifying areas for improvement, and guiding decision-making towards more sustainable practices:

1. **EMS implementation:** Monitor the progress of your organization's environmental management system (EMS) implementation, such as the percentage of facilities covered by the EMS or the number of employees trained in EMS principles.

2. **ISO 14001 certification:** Track the number or percentage of your organization's facilities that have achieved ISO 14001 certification or another recognized EMS standard.

3. **Environmental objectives and targets:** Measure the progress towards achieving your organization's environmental objectives and targets, such as reductions in

energy consumption, waste generation, or greenhouse gas emissions.

4. **Environmental compliance:** Monitor the number of environmental incidents, violations, or regulatory enforcement actions related to your organization's operations, and track improvements in compliance over time.

5. **Environmental audits:** Track the number and frequency of internal or external environmental audits conducted at your organization's facilities, and assess the findings and recommendations for improvement.

6. **Corrective and preventive actions:** Measure the number of corrective and preventive actions (CAPAs) implemented as a result of environmental audits, incidents, or non-compliance, and monitor their effectiveness in addressing the identified issues.

7. **Continuous improvement:** Assess the improvements made in your organization's environmental performance over time, such as reductions in energy consumption, waste generation, or emissions.

8. **Stakeholder engagement:** Track the level and effectiveness of your organization's engagement with stakeholders on environmental issues, such as employee training, community outreach, or partnerships with NGOs and regulators.

9. **Environmental risk management:** Monitor the effectiveness of your organization's environmental risk management processes, such as the identification, assessment, and mitigation of environmental risks associated with your operations, products, or services.

10. **EMS integration:** Assess the degree of integration between your organization's EMS and other management

systems, such as quality management or occupational health and safety management, to promote a comprehensive and coordinated approach to sustainability.

Technology for Environmental Sustainability

Embracing technology for environmental sustainability is crucial for minimizing your environmental impact, enhancing efficiency, and supporting sustainable development. By leveraging innovative technologies, you can improve your environmental performance, reduce costs, and demonstrate your commitment to sustainability. Here are some strategies and practical examples to help you integrate technology for environmental sustainability into your organization:

- **Energy-efficient technologies:** Invest in energy-efficient equipment, appliances, and systems to reduce your energy consumption and greenhouse gas emissions.

  Practical example: Upgrade your office lighting to energy-efficient LED bulbs or install smart thermostats to optimize heating and cooling, reducing energy consumption and costs.

- **Renewable energy technologies:** Implement renewable energy solutions, such as solar panels, wind turbines, or geothermal systems, to generate clean, renewable energy for your operations.

  Practical example: Install solar panels on your facility's rooftop to generate electricity, reducing your reliance on fossil fuels and lowering your carbon footprint.

- **Waste reduction technologies:** Utilize technology to minimize waste generation, improve recycling rates, and support a circular economy.

  Practical example: Implement a digital document management system to reduce paper consumption, or use

3D printing to create prototypes and minimize material waste in product development.

- **Water-saving technologies:** Implement water-efficient technologies and processes to conserve water resources and reduce your water footprint.

  Practical example: Install low-flow faucets and fixtures in your facilities or use smart irrigation systems to optimize water use in landscaping and agriculture.

- **Pollution control technologies:** Adopt technologies that help prevent and control pollution, such as air filtration systems, wastewater treatment technologies, or noise reduction solutions.

  Practical example: Install advanced air filtration systems in your manufacturing facilities to capture and remove pollutants, ensuring clean air emissions and protecting local air quality.

- **Environmental monitoring technologies:** Use sensors, remote sensing, and data analytics to monitor your environmental performance, track resource consumption, and identify opportunities for improvement.

  Practical example: Deploy IoT-enabled sensors to monitor energy and water usage in real-time, providing data to identify inefficiencies and optimize resource consumption.

- **Clean transportation technologies:** Invest in clean transportation solutions, such as electric vehicles, hydrogen fuel cells, or biofuels, to reduce emissions from your fleet and support sustainable mobility.

  Practical example: Replace your company's gasoline-powered vehicles with electric or hybrid models and provide charging infrastructure to support their use.

- **Digital collaboration tools:** Leverage digital technologies to reduce the need for business travel and in-person meetings, minimizing your organization's carbon footprint.

  Practical example: Implement video conferencing platforms and collaboration tools to enable remote work and virtual meetings, reducing emissions associated with commuting and business travel.

Integrating technology for environmental sustainability into your operations will allow you to improve your environmental performance, reduce costs, and demonstrate your commitment to sustainable business practices. This can enhance your reputation, attract customers and investors who value sustainability, and contribute to a more sustainable future.

Module 2 Exam Questions

1. What is the primary purpose of an Environmental Management System (EMS)?

a) To improve employee morale

b) To manage an organization's environmental performance

c) To increase profits

d) To reduce costs

2. Which of the following is a renewable energy source?

a) Solar power

b) Coal

c) Natural gas

d) Petroleum

3. What is an example of a waste reduction technology?

a) Wind turbine

b) Geothermal system

c) Digital document management system

d) Wastewater treatment plant

4. Which of the following is a strategy for reducing greenhouse gas emissions?

a) Purchasing additional gasoline-powered vehicles

b) Expanding office spaces

c) Increasing energy efficiency

d) Investing in non-renewable energy sources

5. What is the main goal of sustainable procurement?

a) To increase profits

b) To prioritize environmental, social, and economic factors

c) To reduce costs

d) To improve employee morale

6. Which of the following is an example of a sustainable transportation option?

a) Carpooling

b) Taking a cruise ship

c) Flying

d) Driving a gasoline-powered car

7. What type of sustainability does ISO 14001 focus on?

a) Environmental sustainability

b) Economic sustainability

c) Governance sustainability

d) Social sustainability

8. Which technology can help conserve water resources?

a) LED lighting

b) Solar panels

c) Low-flow faucets and fixtures

d) Air filtration systems

9. What is an example of a pollution control technology?

a) Solar panel

b) Electric vehicle

c) Video conferencing platform

d) Air filtration system

10. What is the purpose of a legal register in an EMS?

a) To track employee performance

b) To list all relevant environmental laws and regulations

c) To store company policies

d) To record financial transactions

11. How can an organization reduce its carbon footprint in its supply chain?

a) Partnering with suppliers that prioritize non-renewable energy

b) Sourcing raw materials from certified sustainable sources

c) Using excessive packaging materials

d) Ignoring energy efficiency

12. What does a SMART environmental objective stand for?

a) Simple, Manageable, Adaptable, Reliable, Transparent

b) Specific, Measurable, Achievable, Relevant, Time-bound

c) Strategic, Measurable, Accountable, Resilient, Timely

d) Speedy, Methodical, Accessible, Realistic, Tangible

13. Which of the following is an example of a sustainable sourcing strategy?

a) Using single-use plastics whenever possible

b) Purchasing products with excessive packaging

c) Ignoring certifications like Fair Trade or Rainforest Alliance

d) Prioritizing products made from recycled materials

14. What is the main goal of an energy-efficient technology?

a) To minimize waste generation

b) To improve air quality

c) To increase water usage

d) To reduce energy consumption

15. Which of the following is a clean transportation technology?

a) Gasoline-powered vehicles

b) Jet planes

c) Diesel trucks

d) Electric vehicles

16. How can you encourage sustainable transportation among your employees?

a) Discourage public transit use

b) Offer free parking for single-occupancy vehicles

c) Ignore alternative transportation options

d) Promote carpooling

17. What is the purpose of a carbon offset project?

a) To promote waste generation

b) To increase the use of non-renewable resources

c) To reduce greenhouse gas emissions elsewhere

d) To ignore environmental concerns

18. Which of the following is a strategy for reducing water consumption?

a) Overwatering plants and lawns

b) Increasing the use of single-use plastics

c) Leaving the water running while washing hands

d) Installing low-flow faucets and fixtures

19. What is a key aspect of a successful environmental management system?

a) Ignoring environmental regulations

b) Prioritizing short-term gains

c) Focusing solely on financial performance

d) Continuous improvement

20. How can an organization promote biodiversity and ecosystem preservation?

a) Implementing habitat conservation projects

b) Encouraging invasive species propagation

c) Using pesticides indiscriminately

d) Contributing to deforestation

Answers:

1. b
2. a
3. c
4. c
5. b
6. a
7. a
8. c
9. d
10. b
11. b
12. b
13. d
14. d
15. d
16. d
17. c
18. d
19. d
20. a

Module 3. Social Sustainability

Introduction to Social Sustainability

Welcome to Module 3 on Social Sustainability, an essential component of the overall ESG framework. This module will provide you with a comprehensive understanding of social sustainability and its importance in today's business landscape. As you progress through this module, you will learn about various aspects of social sustainability, including human rights, labor standards, employee well-being, diversity, equity, and inclusion, community engagement, and product responsibility. By the end of this module, you will have a solid foundation for incorporating social sustainability into your business strategy and practices.

Social sustainability focuses on the impact of a business on its employees, customers, communities, and other stakeholders. It emphasizes the importance of promoting social equity, inclusiveness, and well-being while ensuring long-term business success. By adopting socially responsible practices, companies can create a positive work environment, enhance customer trust, and contribute to the sustainable development of communities in which they operate.

Throughout this module, we will delve into the key elements of social sustainability and provide practical examples of how businesses can address these issues. From ensuring fair labor practices and upholding human rights across the entire value chain to fostering a diverse and inclusive workplace culture, we will explore the various ways in which organizations can make a difference in the lives of their stakeholders.

Furthermore, this module will introduce you to the critical role that businesses play in the communities they operate in, from philanthropic initiatives to support for education and healthcare. You will also learn about the importance of product responsibility and customer safety as essential aspects of social sustainability.

As you navigate through this module, you will gain valuable insights into the various dimensions of social sustainability and how to integrate them into your organization's strategy and operations. By prioritizing social sustainability, businesses can build a strong foundation for long-term success while contributing to a more just and equitable society.

So, let's embark on this journey together and explore the fascinating world of social sustainability. We hope you enjoy the module and find it insightful and inspiring!

Human Rights and Labor Standards

Human rights and labor standards are fundamental aspects of social sustainability. They ensure that your business respects and promotes the rights and well-being of all individuals involved in your operations, including employees, contractors, and suppliers. In this section, we will discuss the importance of upholding human rights and labor standards, along with practical examples to help you integrate these principles into your organization.

Human rights, as defined by the United Nations, include a wide range of rights, such as the right to life, liberty, and security, freedom from torture and slavery, and the right to work and education. To respect and promote human rights in your business, it is essential to be aware of the potential risks and challenges that your operations may pose to these rights. This may include ensuring fair wages, preventing child labor, and guaranteeing freedom of association for your employees.

Labor standards, on the other hand, are the norms and regulations that govern working conditions and employee rights. These standards encompass aspects such as working hours, occupational health and safety, and discrimination in the workplace. By adhering to international labor standards, such as those established by the International Labour Organization (ILO), you can create a safe, fair, and inclusive work environment for your employees.

To incorporate human rights and labor standards into your business practices, consider the following practical examples:

- **Conduct a thorough human rights impact assessment:** Identify the potential risks and adverse impacts that your operations may have on human rights. This will help you prioritize areas for improvement and develop a targeted action plan.

- **Develop and implement a human rights policy:** Establish a clear policy that outlines your commitment to respecting and promoting human rights throughout your organization. Ensure that this policy is communicated to all employees, suppliers, and other stakeholders.

- **Collaborate with suppliers and partners:** Engage with your suppliers and business partners to ensure they share your commitment to human rights and labor standards. Implement supplier code of conduct, regular audits, and capacity-building programs to help them improve their practices.

- **Train and educate your employees:** Provide ongoing training and education to your employees on human rights and labor standards. This will help them understand their rights and responsibilities and empower them to contribute to a more socially sustainable workplace.

- **Monitor and report on progress:** Regularly assess your performance in upholding human rights and labor standards. Transparently report your progress to stakeholders and use their feedback to continuously improve your practices.

Focusing on human rights and labor standards will allow you to not only protect and empower your employees but also strengthen your reputation and relationships with customers, investors, and other stakeholders. This commitment to social sustainability will ultimately contribute to the long-term success and resilience of your organization.

Practical KPIs for Human Rights and Labor Standards

To effectively measure your organization's performance in relation to human rights and labor standards, you can use the following Key Performance Indicators (KPIs):

1. **Percentage of suppliers assessed for human rights and labor standards:** This KPI measures the proportion of your supply chain partners that have been assessed for their adherence to human rights and labor standards.

2. **Number of reported incidents related to human rights and labor standards:** Track the number of incidents, such as forced labor, child labor, or discrimination, reported within your organization or supply chain.

3. **Percentage of employees receiving human rights and labor standards training:** Measure the proportion of your workforce that has received training on human rights and labor standards, ensuring they are aware of their rights and responsibilities.

4. **Gender pay gap:** Calculate the difference between the average earnings of male and female employees to assess any potential wage discrepancies.

5. **Employee turnover rate:** Monitor the rate at which employees leave your organization, as high turnover can indicate potential issues with labor standards or overall employee satisfaction.

6. **Health and safety incident rate:** Track the number of workplace accidents or incidents per a specific number of worked hours (e.g., per 100,000 hours worked) to assess the safety of your working environment.

7. **Freedom of association and collective bargaining coverage:** Measure the percentage of your workforce covered by collective bargaining agreements, ensuring employees have the right to negotiate their working conditions.

8. **Percentage of workforce covered by formal grievance mechanisms:** Calculate the proportion of employees who

have access to formal channels for raising concerns or complaints related to human rights and labor standards.

Employee Health, Safety, and Well-being

Employee health, safety, and well-being are essential aspects of social sustainability and contribute significantly to the overall success of your business. By prioritizing these areas, you can create a productive, positive, and supportive work environment that fosters employee satisfaction and loyalty. In this section, we will discuss the importance of employee health, safety, and well-being, along with practical examples to help you integrate these principles into your organization.

Employee health and safety involve ensuring that your workplace is free from hazards and risks that could lead to accidents, injuries, or illnesses. By implementing robust health and safety policies, procedures, and training programs, you can minimize the risk of incidents and create a safer work environment for your employees.

Employee well-being encompasses the physical, mental, and emotional aspects of an individual's work experience. By promoting well-being, you can help your employees thrive, both professionally and personally, while also boosting their engagement, productivity, and overall satisfaction.

To prioritize employee health, safety, and well-being in your organization, consider the following practical examples:

- **Develop and implement a comprehensive health and safety management system:** Establish policies, procedures, and guidelines that address potential hazards and risks in your workplace. Ensure that these policies are communicated to all employees and are regularly reviewed and updated.

- **Provide ongoing training and education:** Offer training and education programs to your employees on topics such as occupational health and safety, stress management, and work-life balance. This will help them become more aware of potential risks and better equipped to manage their own well-being.

- **Encourage open communication and feedback:** Foster a culture of open communication where your employees feel comfortable discussing health, safety, and well-being concerns. Regularly solicit feedback from your employees and use their input to improve your practices.

- **Offer flexible work arrangements:** Consider providing flexible work options, such as remote work, flexible hours, or job-sharing, to help your employees achieve a better work-life balance and reduce stress.

- **Create a supportive work environment:** Develop initiatives and programs that promote a positive workplace culture, such as employee recognition programs, team-building activities, and mental health resources.

- **Promote physical well-being:** Encourage your employees to stay active and healthy by offering fitness-related benefits, such as gym memberships, wellness programs, or access to onsite exercise facilities.

Investing in your employees' health, safety, and well-being will allow you to create a more engaged and productive workforce, reduce absenteeism and turnover, and enhance your organization's reputation as an employer of choice. Ultimately, prioritizing these areas will contribute to a more socially sustainable and successful business.

Practical KPIs Employee Health, Safety, and Well-being

To effectively measure your organization's performance in relation to employee health, safety, and well-being, you can use the following Key Performance Indicators (KPIs):

1. **Lost Time Injury Frequency Rate (LTIFR):** Calculate the number of lost time injuries (work-related injuries that result in time away from work) per a specific number of worked hours (e.g., per million hours worked) to assess the safety of your working environment.

2. **Total Recordable Incident Rate (TRIR):** Measure the number of work-related injuries and illnesses per a specific number of worked hours (e.g., per 100,000 hours worked) to track the overall safety performance of your organization.

3. **Employee absenteeism rate:** Monitor the percentage of unscheduled absences in your organization, as high absenteeism can indicate potential issues with employee well-being or working conditions.

4. **Employee satisfaction and well-being survey results:** Conduct regular surveys to assess employee satisfaction and well-being, tracking key indicators such as job satisfaction, stress levels, and work-life balance.

5. **Percentage of employees participating in health and well-being programs:** Measure the proportion of your workforce that actively participates in company-sponsored health and well-being initiatives, such as fitness programs, mental health resources, or wellness workshops.

6. **Ergonomic assessment coverage:** Calculate the percentage of employees who have received an ergonomic assessment, ensuring that their workstations and working conditions are optimized for comfort and safety.

7. **Health and safety training completion rate:** Track the percentage of employees who have completed mandatory health and safety training courses, ensuring they are aware of potential risks and best practices for maintaining a safe working environment.

8. **Number of near-miss incidents reported:** Monitor the number of near-miss incidents (situations where an injury or accident was narrowly avoided) reported within your organization, as they can serve as an early warning sign for potential safety issues.

Diversity, Equity, and Inclusion

Diversity, equity, and inclusion are vital components of social sustainability and play a crucial role in your organization's long-term success. Embracing diversity and fostering an inclusive work environment can lead to increased creativity, innovation, and employee satisfaction. In this section, we will discuss the importance of diversity, equity, and inclusion, along with practical examples to help you integrate these principles into your organization.

Diversity refers to the variety of differences among people, including but not limited to, race, gender, age, ethnicity, sexual orientation, and physical abilities. By valuing and promoting diversity, you can ensure that your organization benefits from a broad range of perspectives, ideas, and experiences.

Equity is about ensuring that all individuals have equal access to opportunities and resources, and are treated fairly and respectfully. This involves addressing any barriers that may exist within your organization and working to create a level playing field for all employees.

Inclusion focuses on creating an environment where everyone feels valued, respected, and supported. This means fostering a culture of belonging, where diverse perspectives are not only recognized but actively sought and integrated into decision-making processes.

To foster diversity, equity, and inclusion in your organization, consider the following practical examples:

- **Develop and implement a diversity, equity, and inclusion strategy:** Establish a clear vision, goals, and action plan for promoting diversity, equity, and inclusion within your organization. Ensure that this strategy is communicated to all employees and integrated into your overall business objectives.

- **Review and update your recruitment and hiring practices:** Ensure that your recruitment and hiring processes are free from bias and provide equal opportunities for candidates from diverse backgrounds. This might involve using blind recruitment techniques, partnering with organizations that focus on diversity in the workforce, or setting diversity targets for your hiring teams.

- **Offer diversity and inclusion training:** Provide ongoing training and education for your employees on topics related to diversity, equity, and inclusion. This will help them become more aware of their own biases, develop cultural competence, and learn how to foster an inclusive work environment.

- **Establish employee resource groups:** Create affinity groups within your organization, such as groups for women, people of color, or LGBTQ+ employees. These groups can provide support, networking opportunities, and a platform for sharing ideas and experiences.

- **Implement inclusive leadership practices:** Encourage leaders within your organization to actively promote diversity, equity, and inclusion. This might involve seeking diverse perspectives in decision-making processes, recognizing and addressing unconscious biases, or acting as mentors and sponsors for underrepresented employees.

- **Measure and report on progress:** Regularly assess your organization's performance in promoting diversity, equity, and inclusion. Transparently report your progress to stakeholders and use their feedback to continuously improve your practices.

Prioritizing diversity, equity, and inclusion will allow you to create a more engaged, innovative, and resilient workforce. This commitment to social sustainability will not only contribute to the long-term success of your organization but also help create a more just and equitable society.

Practical KPIs for Diversity, Equity, and Inclusion

The following KPIs can be used to track/measure progress in promoting diversity, equity, and inclusion (DEI) in the workplace, helping to identify areas for improvement, assess the effectiveness of policies and programs, and guide decision-making towards a more inclusive and equitable work environment:

To effectively measure your organization's performance in relation to diversity, equity, and inclusion, you can use the following Key Performance Indicators (KPIs):

1. **Workforce diversity metrics:** Monitor the demographic composition of your workforce, tracking indicators such as gender, age, ethnicity, and disability status. This will help you assess the diversity of your organization and identify areas for improvement.

2. **Diversity in leadership roles:** Calculate the percentage of underrepresented groups in management and executive positions to evaluate the inclusivity of your organization's leadership.

3. **Pay equity ratio:** Compare the average earnings of different demographic groups within your organization to assess potential wage discrepancies and ensure equal pay for equal work.

4. **Inclusion index:** Use employee surveys to assess the overall level of inclusion in your organization, tracking indicators such as a sense of belonging, fair treatment, and equal opportunity for advancement.

5. **Employee retention rate by demographic group:** Monitor the retention rates of different demographic groups within your organization to identify potential disparities and ensure that all employees feel valued and supported.

6. **Participation in diversity and inclusion training:** Measure the percentage of employees who have completed diversity and inclusion training programs, ensuring they are aware of the importance of these issues and best practices for promoting an inclusive work environment.

7. **Number of diversity and inclusion initiatives:** Track the number of initiatives your organization has implemented to promote diversity, equity, and inclusion, such as mentorship programs, employee resource groups, or recruitment partnerships with diverse organizations.

8. **Employee referrals from underrepresented groups:** Monitor the percentage of employee referrals that come from underrepresented demographic groups to assess the effectiveness of your efforts to create a more diverse and inclusive workplace.

Community Engagement and Development

Community engagement and development are crucial aspects of social sustainability that involve building strong, mutually beneficial relationships with the communities in which your business operates. By actively engaging with and supporting local communities, you can create shared value, improve your reputation, and contribute to the overall well-being and development of the community. In this section, we will discuss the importance of community engagement and development, along with practical examples to help you integrate these principles into your organization.

Community engagement refers to the process of actively involving community members in your business activities, decision-making processes, and initiatives. This might involve seeking their input, sharing information, or collaborating on projects that address local needs and priorities.

Community development focuses on supporting and empowering communities to improve their overall well-being and quality of life. This might involve investing in education, healthcare, infrastructure, or other initiatives that contribute to the long-term social, economic, and environmental sustainability of the community.

To foster community engagement and development within your organization, consider the following practical examples:

- **Conduct a community needs assessment:** Identify the specific needs, priorities, and challenges facing the communities in which you operate. This will help you better understand their concerns and develop targeted initiatives that address these issues.

- **Develop and implement a community engagement strategy:** Establish a clear vision, goals, and action plan for engaging with and supporting your local communities. Ensure that this strategy is communicated to all employees and integrated into your overall business objectives.

- **Create partnerships with local organizations:** Collaborate with local nonprofits, educational institutions, and government agencies to develop and implement community development initiatives. These partnerships can help you leverage resources, expertise, and networks to create more significant impact.

- **Encourage employee volunteerism:** Offer your employees opportunities to volunteer their time and skills to support community initiatives. This might involve organizing company-wide volunteer days, offering paid time off for volunteer activities, or matching employee donations to local causes.

- **Support local businesses and suppliers:** Consider sourcing goods and services from local businesses and suppliers to support economic development within your community. This can help create jobs, strengthen local supply chains, and contribute to the overall sustainability of the community.

- **Measure and report on progress:** Regularly assess your organization's performance in engaging with and supporting local communities. Transparently report your progress to stakeholders and use their feedback to continuously improve your practices.

Prioritizing community engagement and development will allow you to create lasting, positive impacts on the communities in which you operate. This commitment to social sustainability will not only enhance your organization's reputation and relationships with stakeholders but also contribute to a more just and equitable society.

Practical KPIs for Community Engagement and Development

To effectively measure your organization's performance in relation to community engagement and development, you can use the following Key Performance Indicators (KPIs):

1. **Community investment:** Track the total amount of financial and in-kind contributions your organization makes to local communities, such as donations, sponsorships, or pro-bono services.

2. **Employee volunteer participation rate:** Measure the percentage of employees who participate in company-sponsored volunteer activities or community initiatives, demonstrating your organization's commitment to social responsibility.

3. **Number of community partnerships:** Monitor the number of partnerships your organization has established with local community organizations, NGOs, or educational institutions to collaborate on projects or initiatives that support community development.

4. **Community satisfaction and impact survey results:** Conduct regular surveys to assess the satisfaction and perceived impact of your community engagement initiatives from the perspective of local stakeholders, including residents, partner organizations, and local leaders.

5. **Local hiring and procurement rates:** Calculate the percentage of your workforce and suppliers that come from the local community, demonstrating your organization's commitment to supporting local economic development and reducing its environmental impact.

6. **Number of community development projects supported:** Track the number of community development projects your organization has supported, such as infrastructure improvements, educational programs, or environmental conservation initiatives.

7. **Community outreach and engagement events:** Monitor the number of events or activities your organization organizes or participates in to engage with local communities, such as town hall meetings, community workshops, or open houses.

8. **Impact of community development initiatives:** Assess the tangible outcomes and long-term impact of your community development initiatives, such as improvements in local education, employment, or environmental conditions.

Product Responsibility and Customer Safety

Product responsibility and customer safety are essential components of social sustainability that focus on ensuring the products and services you offer meet the highest standards of quality, safety, and ethical practices. By prioritizing these aspects, you can build trust and loyalty among your customers, reduce the risk of negative impacts on people and the environment, and contribute to a more sustainable society. In this section, we will discuss the importance of product responsibility and customer safety, along with practical examples to help you integrate these principles into your organization.

Product responsibility encompasses the entire lifecycle of your products or services, from design and development to production, distribution, use, and end-of-life management. This involves ensuring that your products are safe, environmentally friendly, and ethically produced, as well as meeting or exceeding relevant industry standards and regulations.

Customer safety focuses on minimizing the risk of harm to your customers when using your products or services. This includes providing clear instructions, warnings, and support to help customers use your products safely and effectively.

To promote product responsibility and customer safety within your organization, consider the following practical examples:

- **Integrate sustainability into product design and development:** Consider the social and environmental impacts of your products during the design and development stages. This might involve using eco-friendly materials, designing for durability and longevity, or considering end-of-life management and recycling options.

- **Implement robust quality management systems:** Develop and maintain quality management systems that ensure your products meet the highest standards of safety, performance, and reliability. This might involve regular testing, inspections, and audits, as well as continuous improvement efforts.
- **Comply with industry standards and regulations:** Ensure your products and services meet or exceed relevant industry standards and regulations related to safety, quality, and environmental performance.
- **Communicate product safety information clearly:** Provide clear and accessible instructions, warnings, and safety information to your customers. This might involve using easy-to-understand language, visual aids, or even offering dedicated customer support to address safety concerns.
- **Monitor and address product safety incidents:** Establish processes for monitoring and addressing any product safety incidents that may occur. This might involve recalling products, offering refunds or replacements, or providing additional support and resources to affected customers.
- **Engage with stakeholders on product responsibility:** Communicate your commitment to product responsibility and customer safety to your stakeholders, including customers, suppliers, and investors. Seek their feedback and input to continuously improve your practices.

By focusing on product responsibility and customer safety, you can create value for your customers, minimize the risk of negative impacts on people and the environment, and contribute to a more sustainable society. This commitment to social sustainability will ultimately enhance your organization's reputation and long-term success.

Practical KPIs for Product Responsibility and Customer Safety

To effectively measure your organization's performance in relation to product responsibility and customer safety, you can use the following Key Performance Indicators (KPIs):

1. **Number of product recalls:** Monitor the number of product recalls due to safety concerns or defects, which can help you identify potential issues with your products and take corrective actions.

2. **Customer satisfaction with product safety:** Conduct regular surveys to assess customer satisfaction with the safety and quality of your products, helping you understand their needs and expectations.

3. **Number of customer complaints related to product safety:** Track the number of customer complaints regarding product safety, which can serve as an early warning sign for potential issues.

4. **Percentage of products meeting safety standards:** Calculate the proportion of your products that meet or exceed industry safety standards and regulatory requirements, demonstrating your commitment to customer safety.

5. **Product safety training completion rate:** Measure the percentage of employees who have completed mandatory product safety training courses, ensuring they are aware of potential risks and best practices for maintaining a safe product portfolio.

6. **Number of safety incidents related to product use:** Monitor the number of safety incidents or accidents that occur during the use of your products, helping you identify potential hazards and improve product safety.

7. **Product lifecycle assessment coverage:** Track the percentage of your products that have undergone a comprehensive lifecycle assessment, evaluating the environmental, social, and safety impacts throughout their lifecycle.

8. **Responsible sourcing and supply chain transparency:** Monitor the proportion of your suppliers and raw materials that meet your organization's standards for responsible sourcing, ethical practices, and safety compliance.

Talent Attraction, Retention, and Development

Talent attraction, retention, and development are critical aspects of social sustainability that involve finding, nurturing, and retaining the best employees for your organization. By focusing on these areas, you can build a skilled, engaged, and diverse workforce that drives innovation, productivity, and long-term success. In this section, we will discuss the importance of talent attraction, retention, and development, along with practical examples to help you integrate these principles into your organization.

Talent attraction involves identifying and recruiting the most talented individuals for your organization. This might involve developing a compelling employer brand, leveraging networks and partnerships, or using innovative recruitment strategies to reach a diverse pool of candidates.

Talent retention is about keeping your best employees engaged and committed to your organization. This involves creating a positive work environment, offering competitive compensation and benefits, and providing opportunities for growth and advancement.

Talent development focuses on investing in the ongoing learning and growth of your employees, helping them acquire new skills, enhance their performance, and achieve their career goals.

To prioritize talent attraction, retention, and development in your organization, consider the following practical examples:

- **Develop a compelling employer brand:** Communicate your organization's values, culture, and commitment to social sustainability to attract like-minded candidates who are passionate about making a positive impact.

- **Leverage diverse recruitment channels:** Use a variety of recruitment channels, such as social media, job fairs, and partnerships with universities or professional associations, to reach a diverse pool of potential candidates.

- **Offer competitive compensation and benefits:** Ensure your compensation and benefits packages are competitive within your industry and aligned with the needs and preferences of your employees. This might involve offering flexible work arrangements, comprehensive health and wellness programs, or generous parental leave policies.

- **Create a supportive work environment:** Foster a positive work culture that values diversity, equity, and inclusion, and promotes employee well-being, engagement, and satisfaction.

- **Provide opportunities for learning and development:** Offer ongoing training, education, and development programs to help your employees grow professionally and personally. This might involve providing access to online courses, offering mentorship programs, or supporting employees in pursuing further education or certifications.

- **Recognize and reward performance:** Develop performance management systems that recognize and reward employee contributions, achievements, and growth. This might involve offering promotions, bonuses, or other forms of recognition that demonstrate your appreciation for their hard work and dedication.

Focusing on talent attraction, retention, and development, will allow you to build a skilled, engaged, and diverse workforce that drives your organization's long-term success. This commitment to social sustainability will not only enhance your reputation as an employer of choice but also contribute to a more just and equitable society.

Practical KPIs for Talent Attraction, Retention, and Development

To effectively measure your organization's performance in relation to talent attraction, retention, and development, you can use the following Key Performance Indicators (KPIs):

1. **Employee turnover rate:** Monitor the rate at which employees leave your organization, as high turnover can indicate potential issues with employee satisfaction, engagement, or development opportunities.

2. **Time-to-fill open positions:** Track the average time it takes to fill open positions within your organization, which can help you assess the effectiveness of your talent attraction strategies.

3. **Employee satisfaction with career development opportunities:** Conduct regular surveys to assess employee satisfaction with the career development opportunities available within your organization, such as training, mentorship, or internal mobility programs.

4. **Percentage of internal promotions:** Calculate the proportion of job vacancies filled by internal candidates, which can indicate the effectiveness of your talent development and retention strategies.

5. **Retention rate of high-potential employees:** Measure the retention rate of employees identified as high-potential or high-performing, ensuring that you are effectively nurturing and retaining top talent.

6. **Employee skill development:** Monitor the number of training hours or professional development courses completed by your employees, demonstrating your organization's commitment to continuous learning and development.

7. **Diversity in new hires:** Track the demographic composition of new hires to ensure that your talent attraction strategies are promoting diversity and inclusion.

8. **Employee engagement survey results:** Use employee engagement surveys to assess key factors such as job satisfaction, work-life balance, and the perceived quality of management, which can help you identify areas for improvement in your talent management practices.

Technology for Social Sustainability

Technology plays a crucial role in driving social sustainability by enabling you to improve various aspects of your organization's performance. By adopting innovative technological solutions, you can enhance your social sustainability efforts, streamline operations, and create positive change. Here are some practical examples of how technology can contribute to social sustainability:

- **Remote work and collaboration tools:** By embracing remote work and collaboration tools, you can create a more flexible work environment that accommodates diverse needs, reduces commuting, and improves work-life balance for your employees.

- **E-learning platforms:** Investing in e-learning platforms allows you to provide continuous learning and development opportunities for your employees, helping them expand their skill sets, enhance their career prospects, and stay engaged with your organization.

- **Data analytics and artificial intelligence:** Leveraging data analytics and AI solutions can help you gain valuable insights into employee satisfaction, well-being, and performance, enabling you to identify trends, address issues, and make data-driven decisions to improve your social sustainability efforts.

- **Virtual reality (VR) and augmented reality (AR) for training:** Using VR and AR technologies, you can create immersive and interactive training experiences, making it easier for your employees to learn and retain new skills, ultimately improving their performance and job satisfaction.

- **Social media and engagement platforms:** Utilizing social media and other digital engagement platforms allows you to connect with your stakeholders, including customers, employees, and local communities, fostering open communication, transparency, and a sense of belonging.

- **Mobile applications for employee well-being:** By offering mobile apps that focus on employee well-being, such as mental health support or fitness tracking, you can encourage your employees to prioritize their health and overall well-being.
- **Blockchain for supply chain transparency:** Implementing blockchain technology in your supply chain can increase transparency, traceability, and accountability, ensuring ethical practices and responsible sourcing across your operations.

Integrating these technologies into your organization will allow you to drive social sustainability by creating a more inclusive, equitable, and supportive environment for your employees, customers, and communities. Embracing technology can be instrumental in achieving your social sustainability goals and fostering long-term, positive change.

Module 3 Exam Questions

1. What does ESG stand for?

a) Economic, Social, Governance

b) Environmental, Social, Governance

c) Environmental, Sustainable, Growth

d) Ecological, Social, Growth

2. Which of the following is an example of an environmental sustainability initiative?

a) Reducing greenhouse gas emissions

b) Promoting diversity and inclusion

c) Ensuring ethical supply chain practices

d) Engaging with local communities

3. What is the primary focus of social sustainability?

a) Protecting the environment

b) Ensuring ethical corporate governance

c) Fostering positive relationships with stakeholders

d) Minimizing financial

4. Which aspect of governance sustainability deals with executive compensation?

a) Board composition and diversity

b) Shareholder rights and engagement

c) Executive compensation and performance management

d) Business ethics and anti-corruption

5. What is a key element of community engagement and development?

a) Reducing greenhouse gas emissions

b) Monitoring product safety

c) Supporting local economic development

d) Ensuring regulatory compliance

6. Which of these is a practical KPI for diversity, equity, and inclusion?

a) Employee turnover rate

b) Percentage of diverse employees in management positions

c) Number of product recalls

d) Local hiring rate

7. What is the main benefit of using remote work and collaboration tools for social sustainability?

a) Reducing greenhouse gas emissions

b) Enhancing employee work-life balance

c) Ensuring product safety

d) Increasing supply chain transparency

8. Which of the following is NOT a category within the ESG framework?

a) Environmental sustainability

b) Social sustainability

c) Economic sustainability

d) Governance sustainability

9. What is the purpose of ESG reporting and disclosure?

a) To attract top talent

b) To communicate ESG performance and transparency to stakeholders

c) To improve product quality

d) To comply with environmental regulations

10. Which of the following factors is considered in responsible investing?

a) ESG factors

b) Company size

c) Market trends

d) None of the above

11. How can technology contribute to social sustainability?

a) By enhancing employee training

b) By improving employee well-being

c) By increasing supply chain transparency

d) All of the above

12. Which of the following KPIs is related to employee health, safety, and well-being?

a) Employee turnover rate

b) Time-to-fill open positions

c) Number of safety incidents related to product use

d) None of the above

13. What is a key element of product responsibility and customer safety?

a) Ensuring products meet safety standards

b) Reducing greenhouse gas emissions

c) Fostering diversity and inclusion

d) Engaging with local communities

14. What is the primary goal of environmental sustainability?

a) Ensuring corporate governance

b) Protecting the environment and natural resources

c) Fostering positive relationships with stakeholders

d) Minimizing financial risks

15. Which of the following KPIs is related to community engagement and development?

a) Employee turnover rate

b) Number of community partnerships

c) Percentage of products meeting safety standards

d) Retention rate of high-potential employees

16. How can e-learning platforms contribute to social sustainability?

a) By reducing carbon emissions

b) By improving employee satisfaction and development

c) By increasing supply chain transparency

d) By enhancing product safety

17. Which of the following is a practical KPI for human rights and labor standards?

a) Employee turnover rate

b) Number of child labor incidents in the supply chain

c) Number of safety incidents related to product use

d) Retention rate of high-potential employees

18. What is the purpose of ESG risk and opportunity assessment?

a) To identify and prioritize ESG factors that impact your organization

b) To attract top talent

c) To improve product quality

d) To comply with environmental regulations

19. Which of the following is a practical KPI for talent attraction, retention, and development?

a) Employee turnover rate

b) Number of community partnerships

c) Percentage of products meeting safety standards

d) Number of safety incidents related to product use

20. What is the role of technology and innovation in the future of Sustainability?

a) Reducing environmental impact

b) Enhancing social sustainability initiatives

c) Improving corporate governance

d) All of the above

Answers:

1. b
2. a
3. c
4. c
5. c
6. b
7. b
8. c
9. b
10. a
11. d
12. d
13. a
14. b
15. b
16. b
17. b
18. a
19. a
20. d

Module 4. Governance Sustainability

Introduction to Governance Sustainability

Welcome to Module 4 on Governance Sustainability. In this module, you will learn about the crucial role of strong corporate governance in promoting a company's environmental, social, and financial performance. As organizations strive to achieve long-term success and sustainability, implementing effective governance practices has become more important than ever.

Throughout this module, we will explore the key components of governance sustainability, discussing the relevance of various structures, policies, and practices to the overall ESG framework. By incorporating real-life examples and case studies, we aim to provide you with a comprehensive understanding of how good corporate governance can lead to positive outcomes for businesses and society.

As we progress through this module, you will gain valuable insights into the importance of diverse board compositions, responsible executive compensation structures, and transparent shareholder engagement practices. We will also touch upon the significance of robust business ethics, anti-corruption policies, and effective risk management in maintaining a company's reputation and ensuring long-term success.

By the end of this module, you will have a solid foundation in governance sustainability, which will enable you to make informed decisions and contribute to the development of sustainable business strategies. Remember, good corporate governance is not just about compliance; it is about fostering an ethical and responsible corporate culture that drives continuous improvement and innovation in ESG performance.

Corporate Governance Structures and Policies

As you explore the realm of governance sustainability, understanding corporate governance structures and policies is essential. These structures and policies serve as the backbone for effective management and decision-making within your organization, helping to balance the interests of various stakeholders.

Corporate governance structures typically include the following elements:

Board of Directors: The board is responsible for overseeing your company's strategic direction, ensuring sound management, and protecting the interests of shareholders. To foster good governance, your board should be composed of a mix of executive and non-executive (independent) directors with diverse skills, backgrounds, and experiences.

Board Committees: Committees, such as the audit, remuneration, and nomination committees, are established to focus on specific governance aspects. These committees help to streamline the decision-making process, improve oversight, and maintain accountability within your organization.

Management Team: Led by the CEO, the management team is responsible for executing the company's strategy and day-to-day operations. Ensuring a clear delineation of roles and responsibilities between the board and management team is vital for effective governance.

When it comes to corporate governance policies, your organization should consider implementing the following:

- **Code of Conduct:** This policy outlines the ethical standards and behaviors expected of your board, management, and employees. By adhering to a well-defined code of conduct, you can foster a culture of integrity and accountability within your organization.

- **Conflict of Interest Policy:** This policy guides board members and employees in identifying, disclosing, and managing potential conflicts of interest, ensuring that decisions are made in the best interest of your company.
- **Whistleblower Policy:** A whistleblower policy encourages employees to report unethical or illegal activities, protecting them from retaliation and fostering a transparent and accountable organizational culture.
- **Shareholder Engagement Policy:** This policy outlines your approach to engaging with shareholders and addressing their concerns, promoting transparency and fostering constructive relationships with investors.

Practical Examples:

- **Alphabet Inc.**, the parent company of Google, has implemented a dual-class share structure, which separates voting rights from economic interests. This structure allows the founders to retain control over strategic decisions while still raising capital from shareholders.
- **Unilever**, a multinational consumer goods company, has a clear separation of roles between its Chairman and CEO, ensuring an appropriate balance of power and avoiding the concentration of authority in a single individual.
- **Starbucks** has a comprehensive code of conduct, called the "Starbucks Global Human Rights Standard," which applies to all employees and covers various aspects of ethical conduct, human rights, and environmental sustainability.

As you develop and refine your corporate governance structures and policies, keep in mind that they should be tailored to your organization's specific needs and context. By doing so, you can foster a culture of good governance that drives long-term success and sustainability.

Practical KPIs for Corporate Governance Structures and Policies

To measure the effectiveness of corporate governance structures and policies, you can track the following Key Performance Indicators (KPIs):

1. **Board independence ratio:** Calculate the proportion of independent directors on your board compared to the total number of board members. A higher ratio indicates a more independent and objective board, which is beneficial for effective decision-making and oversight.

2. **Board meeting attendance rate:** Track the attendance rate of board members at board meetings. High attendance rates demonstrate the commitment and engagement of board members in their governance roles.

3. **Board evaluation frequency:** Measure how often your organization conducts board evaluations. Regular evaluations help identify areas for improvement and ensure that your board remains effective and responsive to evolving business needs.

4. **Policy review frequency:** Assess how often your organization reviews and updates its corporate governance policies. Regular reviews ensure that your policies remain relevant and aligned with best practices and regulatory requirements.

5. **Whistleblower reports:** Monitor the number of whistleblower reports received by your organization. This KPI can provide insights into the effectiveness of your governance policies and internal controls, as well as your organization's ethical culture.

6. **Training and development hours for directors:** Track the number of hours spent on training and development programs for board members. This KPI indicates the organization's commitment to enhancing the skills and expertise of its board members.

Board Composition and Diversity

One of the critical aspects of effective corporate governance is a diverse and well-balanced board composition. A diverse board, comprising members with varied backgrounds, skills, and experiences, can significantly contribute to your organization's performance and decision-making processes. By promoting diversity, you can ensure that your board is better equipped to navigate the complexities of today's business environment and address the evolving needs of stakeholders.

There are several dimensions of board diversity to consider:

- **Functional Diversity:** Board members should possess a wide range of expertise, including finance, strategy, operations, legal, human resources, and technology. This functional diversity allows your board to provide well-rounded guidance and oversight to your organization.

- **Gender Diversity:** Increasing gender diversity on boards has been linked to improved financial performance, enhanced innovation, and better decision-making. By including more women on your board, you can ensure a broader range of perspectives and experiences are represented.

- **Ethnic and Cultural Diversity:** A board with members from different ethnic and cultural backgrounds can offer unique insights and perspectives, helping your organization better understand and serve diverse markets and stakeholders.

- **Age and Tenure Diversity:** Including board members with different ages and lengths of service can provide a healthy mix of fresh perspectives and seasoned experience, fostering a dynamic and adaptable board.

Practical Examples:

- **General Motors (GM)** has set a strong example of gender diversity at the board level. As of 2021, GM's board comprised 50% women, including its Chair and CEO, Mary Barra. This diverse board composition has been credited with driving innovation and improving the company's performance.

- **Microsoft**'s board is an excellent example of ethnic and cultural diversity. With members hailing from various parts of the world, including the United States, India, and the United Kingdom, Microsoft's board is well-positioned to understand the global market and make informed decisions.

- **IBM** has embraced age diversity by appointing younger board members who bring fresh perspectives and expertise in emerging technologies, such as artificial intelligence and blockchain, to complement the experience of their more seasoned counterparts.

As you work towards enhancing board composition and diversity in your organization, remember that diversity should be viewed as a valuable asset that can contribute to your company's long-term success and sustainability. By embracing diversity at the highest levels of governance, you can foster a more inclusive and innovative organizational culture.

Practical KPIs for Board Composition and Diversity

To measure the effectiveness of board composition and diversity in your organization, consider tracking the following Key Performance Indicators (KPIs):

1. **Gender diversity ratio:** Calculate the proportion of female directors on your board compared to the total number of board members. A higher ratio indicates better gender diversity and representation.

2. **Ethnic and racial diversity ratio:** Measure the proportion of directors from diverse ethnic and racial backgrounds on your board. A higher ratio reflects a more inclusive and diverse board, which can contribute to better decision-making and performance.

3. **Age diversity:** Assess the age range of your board members, considering factors such as average age, age distribution, and generational representation. A diverse age range can bring different perspectives and experiences to the board.

4. **Board tenure diversity:** Calculate the range of board member tenures, considering factors such as average tenure and the distribution of tenures. A diverse mix of tenure can provide a balance between experienced directors and fresh perspectives.

5. **Industry experience diversity:** Evaluate the range of industry experiences and backgrounds represented on your board. A diverse mix of industry expertise can contribute to more informed decision-making and better risk management.

6. **Educational background diversity:** Assess the variety of educational backgrounds represented on your board, including degrees, disciplines, and institutions. A diverse range of educational backgrounds can bring different perspectives and skills to the board.

Executive Compensation and Performance Management

In the context of governance sustainability, executive compensation and performance management are critical factors in ensuring that your organization remains focused on creating long-term value for all stakeholders. A well-designed compensation structure aligns executive incentives with your company's strategic objectives, performance, and ESG goals.

To create an effective executive compensation and performance management system, consider the following:

- **Align compensation with long-term value creation:** Ensure that your executive compensation structure incentivizes sustainable growth and long-term value creation, rather than short-term gains. This can be achieved by linking a significant portion of compensation to the achievement of strategic and ESG targets.

- **Balance fixed and variable pay:** Your compensation structure should include a mix of fixed and variable pay components. Variable pay, such as bonuses and stock options, should be tied to performance metrics that reflect your organization's financial, environmental, and social goals.

- **Use multiple performance metrics:** Relying on a single performance metric, such as earnings per share, can lead to unintended consequences. Incorporate a diverse set of financial and non-financial metrics in your compensation structure to provide a comprehensive view of performance.

- **Implement clawback and malus provisions:** These provisions allow your company to recover or withhold executive bonuses and incentives in cases of financial restatements, misconduct, or poor performance. They help ensure that executive compensation remains aligned with your organization's long-term interests.

Practical Examples:

- **Unilever** has linked a portion of its executives' long-term incentives to the achievement of its Sustainable Living Plan goals. This approach ensures that the company's top management remains focused on delivering sustainable growth and value creation.

- **Novo Nordisk**, a global healthcare company, uses a balanced scorecard approach to measure executive performance. The scorecard includes financial metrics, such as sales growth and operating profit, as well as non-financial metrics related to environmental and social performance.

- **BP** has implemented a clawback provision in its executive compensation policy, allowing the company to recover bonuses in cases of misconduct, such as the Deepwater Horizon oil spill. This provision has helped to reinforce the importance of safety and risk management in the company's operations.

As you develop and refine your executive compensation and performance management system, ensure that it is designed to promote long-term value creation and sustainability. By aligning executive incentives with your organization's strategic and ESG objectives, you can foster a culture of responsibility and accountability at the highest levels of management.

Practical KPIs for Executive Compensation and Performance Management

To measure the effectiveness of executive compensation and performance management in your organization, consider tracking the following Key Performance Indicators (KPIs):

1. **Pay-for-performance alignment:** Evaluate the correlation between executive compensation and your organization's financial performance, such as revenue growth, profitability, or total shareholder return. A strong alignment indicates that executive pay is effectively tied to organizational performance.

2. **CEO pay ratio:** Calculate the ratio of CEO compensation to the median employee compensation. This KPI provides insights into pay equity within your organization and can be used to assess the fairness of executive pay levels.

3. **Short-term vs. long-term incentives:** Analyze the balance between short-term incentives (such as annual bonuses) and long-term incentives (such as stock options or restricted stock units) in executive compensation packages. A balanced mix of short- and long-term incentives can encourage executives to focus on both immediate and future performance.

4. **Performance metric diversity:** Assess the variety of performance metrics used to determine executive compensation, including financial, operational, and ESG (Environmental, Social, and Governance) metrics. A diverse set of performance metrics can help ensure a more comprehensive and balanced evaluation of executive performance.

5. **Executive retention rate:** Track the retention rate of key executives over time. A high retention rate can indicate that your organization's compensation and performance management practices are effective in retaining top talent.

6. **Goal achievement rate:** Measure the percentage of performance goals achieved by executives. This KPI can help evaluate the effectiveness of your performance management process and the alignment of executive goals with organizational objectives.

Shareholder Rights and Engagement

In the pursuit of governance sustainability, respecting and protecting shareholder rights while actively engaging with them is vital. Shareholder engagement helps ensure that your organization remains accountable, transparent, and responsive to investor concerns, ultimately contributing to long-term success and value creation.

To promote effective shareholder rights and engagement, consider the following practices:

- **Facilitate shareholder voting:** Encourage shareholders to exercise their voting rights by making the process accessible and transparent. This includes providing clear and timely information about shareholder meetings, proxy voting, and the items up for vote.

- **Maintain open lines of communication:** Establish regular communication channels with shareholders, such as investor relations websites, annual reports, and earnings calls. Transparent communication helps build trust and fosters a constructive relationship with investors.

- **Engage in dialogue with shareholders:** Proactively engage with shareholders to discuss their concerns and gather valuable feedback. This can be done through one-on-one meetings, investor conferences, and shareholder advisory panels.

- **Respond to shareholder proposals:** Address shareholder proposals constructively, even if they are non-binding or advisory in nature. Demonstrating a willingness to consider and implement shareholder recommendations can enhance your organization's credibility and reputation.

Practical Examples:

- **Apple** has adopted a "Say on Pay" policy, which allows shareholders to cast an advisory vote on executive compensation. This policy provides shareholders with an opportunity to voice their opinions on the company's compensation practices.
- **Procter & Gamble (P&G)** actively engages with its shareholders through a variety of channels, including an annual shareholder forum where investors can meet with the company's leadership to discuss key issues and concerns.
- In response to shareholder pressure, **Amazon** has committed to greater transparency on its climate-related risks and opportunities. The company now discloses its greenhouse gas emissions and has set ambitious goals to achieve net-zero carbon emissions.

As you work towards enhancing shareholder rights and engagement in your organization, remember that proactive and transparent communication with investors is key. By actively engaging with shareholders and addressing their concerns, you can create a more accountable and resilient organization that is better equipped to deliver long-term value for all stakeholders.

Practical KPIs for Shareholder Rights and Engagement

To measure the effectiveness of shareholder rights and engagement in your organization, consider tracking the following Key Performance Indicators (KPIs):

1. **Shareholder voting participation rate:** Calculate the percentage of eligible shares voted at your organization's annual general meeting (AGM) or other shareholder meetings. A high participation rate indicates strong

shareholder engagement and interest in your organization's governance.

2. **Shareholder proposal rate:** Track the number of shareholder proposals submitted and the percentage of proposals that receive majority support. This KPI can provide insights into shareholder concerns and the responsiveness of your organization's management and board to these concerns.

3. **Shareholder engagement frequency:** Measure the number of formal and informal interactions between your organization's management, board, and shareholders. Regular engagement can help maintain open communication channels, build trust, and promote long-term shareholder relationships.

4. **Shareholder satisfaction rate:** Assess shareholder satisfaction through surveys, feedback forms, or other methods. This KPI can help you gauge the effectiveness of your shareholder engagement efforts and identify areas for improvement.

5. **Proxy advisory firm recommendations:** Monitor the recommendations made by proxy advisory firms, such as Institutional Shareholder Services (ISS) or Glass Lewis, regarding your organization's governance practices. Positive recommendations can indicate that your organization's governance practices are aligned with shareholder interests and industry best practices.

6. **Adoption of shareholder rights policies:** Track the implementation of shareholder-friendly policies, such as majority voting, proxy access, or "say-on-pay" votes. These policies can help enhance shareholder rights and demonstrate your organization's commitment to strong governance practices.

Business Ethics and Anti-Corruption

Adhering to strong business ethics and implementing robust anti-corruption policies are critical components of governance sustainability. By fostering a culture of integrity and ethical conduct within your organization, you can minimize the risk of financial and reputational damage and ensure long-term success.

To promote business ethics and combat corruption, consider implementing the following measures:

- **Develop a code of ethics:** A comprehensive code of ethics should outline the values, principles, and expected behaviors for everyone within your organization. Ensure that the code is easily accessible, and provide training to help employees understand and adhere to these ethical guidelines.

- **Establish clear anti-corruption policies:** Develop policies that specifically address bribery, corruption, and other unethical practices. These policies should cover areas such as gifts, hospitality, facilitation payments, and third-party due diligence.

- **Encourage reporting of ethical concerns:** Create a safe and confidential channel for employees to report potential ethical violations, such as a whistleblower hotline or an anonymous reporting system. Ensure that your organization has a clear policy in place to protect employees from retaliation.

- **Monitor and enforce compliance:** Regularly assess and audit your organization's adherence to its ethical guidelines and anti-corruption policies. Address any identified issues promptly and ensure that appropriate disciplinary actions are taken when necessary.

Practical Examples:

- **Siemens**, following a significant corruption scandal, implemented a comprehensive compliance program that includes strict anti-bribery and corruption policies, employee training, and a robust internal control system. These measures have helped the company rebuild its reputation and regain trust from stakeholders.
- **Walmart** has established a Global Ethics Office, which is responsible for promoting a culture of integrity within the company. The office provides resources, training, and guidance to employees, as well as a confidential reporting system for ethical concerns.
- **Unilever**'s Partner to Win program is designed to ensure that its suppliers adhere to the company's ethical and sustainability standards. The program includes a comprehensive assessment process and ongoing monitoring of suppliers' performance in areas such as business integrity, human rights, and environmental management.

As you work towards promoting business ethics and combating corruption in your organization, keep in mind that fostering a culture of integrity requires ongoing commitment and vigilance. By implementing robust policies and practices, you can protect your organization's reputation, minimize risk, and create a solid foundation for sustainable success.

Practical KPIs for Business Ethics and Anti-Corruption

To measure the effectiveness of your organization's business ethics and anti-corruption efforts, consider tracking the following Key Performance Indicators (KPIs):

1. **Ethics and compliance training completion rate:** Calculate the percentage of employees who have

completed mandatory ethics and compliance training. A high completion rate indicates a strong commitment to ethical behavior and regulatory compliance across the organization.

2. **Code of conduct acknowledgement rate:** Measure the percentage of employees who have acknowledged and agreed to adhere to your organization's code of conduct. This KPI demonstrates employees' awareness and commitment to ethical business practices.

3. **Number of ethics hotline reports:** Monitor the number of reports received through your organization's ethics hotline or reporting system. This KPI can provide insights into potential ethical issues or areas of concern that may require further investigation or remediation.

4. **Percentage of resolved ethics cases:** Assess the percentage of ethics cases that have been resolved within a specified time frame. A high resolution rate indicates that your organization is effectively addressing and resolving ethical concerns.

5. **Anti-corruption due diligence rate:** Track the percentage of third-party business partners, such as suppliers, contractors, or agents, who have undergone anti-corruption due diligence. This KPI demonstrates your organization's commitment to maintaining an ethical supply chain and mitigating corruption risks.

6. **Number of regulatory violations and fines:** Monitor the number of regulatory violations and fines related to ethical or corruption issues. A low number or downward trend in violations and fines indicates that your organization's ethics and anti-corruption efforts are effective.

Regulatory Compliance and Risk Management

Maintaining regulatory compliance and effective risk management is essential for governance sustainability. By ensuring compliance with relevant laws, regulations, and industry standards, you can protect your organization from legal, financial, and reputational risks, while also demonstrating a commitment to responsible business practices.

To strengthen regulatory compliance and risk management in your organization, consider the following steps:

- **Develop a compliance framework:** Establish a comprehensive compliance framework that encompasses all applicable laws, regulations, and industry standards. This framework should include clear policies, procedures, and controls to help your organization meet its compliance obligations.

- **Implement a risk management process:** Establish a systematic process for identifying, assessing, prioritizing, and mitigating risks across your organization. This process should involve regular risk assessments and the development of risk mitigation strategies and action plans.

- **Allocate resources and responsibilities:** Assign clear roles and responsibilities for compliance and risk management within your organization. Ensure that adequate resources, including personnel, training, and technology, are available to support these functions.

- **Monitor and report on compliance and risk:** Continuously monitor your organization's compliance with relevant regulations and the effectiveness of your risk management process. Regularly report on your progress to your board and other stakeholders.

Practical Examples:

- Pharmaceutical companies, such as **Pfizer** and Merck, must adhere to strict regulatory requirements, including those set by the U.S. Food and Drug Administration (FDA). These companies have robust compliance and risk management programs in place to ensure that their products meet safety, efficacy, and quality standards.
- In the financial services industry, banks like **JPMorgan** Chase and Bank of America must comply with various regulations, such as the Dodd-Frank Wall Street Reform and Consumer Protection Act. These banks have established extensive compliance and risk management frameworks to address areas such as consumer protection, capital adequacy, and anti-money laundering.
- Companies in the energy sector, like **ExxonMobil** and **Chevron**, face complex regulatory requirements related to environmental protection, safety, and sustainability. To maintain compliance, these companies invest in advanced technologies, employee training, and rigorous risk management practices.

As you work to enhance regulatory compliance and risk management within your organization, remember that these efforts should be viewed as an ongoing and proactive process. By staying current with regulatory developments, regularly assessing and managing risks, and fostering a culture of compliance, you can build a resilient and sustainable organization that meets the expectations of regulators, investors, and other stakeholders.

Practical KPIs for Regulatory Compliance and Risk Management

To measure the effectiveness of your organization's regulatory compliance and risk management efforts, consider tracking the following Key Performance Indicators (KPIs):

1. **Number of compliance breaches:** Monitor the number of compliance breaches or violations your organization experiences. A low number or downward trend in breaches indicates effective compliance processes and risk management strategies.

2. **Percentage of resolved compliance issues:** Assess the percentage of identified compliance issues that have been resolved within a specified time frame. A high resolution rate demonstrates your organization's commitment to addressing compliance concerns promptly.

3. **Compliance training completion rate:** Calculate the percentage of employees who have completed mandatory compliance training. A high completion rate indicates a strong commitment to regulatory compliance and employee awareness of relevant rules and regulations.

4. **Risk assessment coverage:** Measure the percentage of key business processes, departments, or functions that have undergone a formal risk assessment. Comprehensive risk assessment coverage can help ensure that your organization is effectively identifying and managing potential risks.

5. **Risk mitigation progress:** Track the progress of risk mitigation actions or initiatives, such as the percentage of planned actions completed or the reduction in risk exposure. This KPI demonstrates your organization's commitment to managing and reducing risks.

6. **Regulatory audit results:** Monitor the results of regulatory audits, such as the number of findings, the severity of issues identified, and any fines or penalties imposed. Positive audit results indicate effective compliance and risk management practices.

7. **Key risk indicators (KRIs):** Identify and track KRIs that are specific to your organization's industry, operations, and

risk profile. Monitoring KRIs can help you detect emerging risks and trends, enabling proactive risk management and mitigation.

Technology for Governance Sustainability

In this section, you will explore the role of technology in enhancing governance sustainability. The integration of technology into governance practices can help streamline processes, improve transparency, and facilitate better decision-making. Here are some ways you can leverage technology for governance sustainability:

Board management software

Utilize board management software to centralize board documents, meeting agendas, and minutes, ensuring easy access for all board members. This software can also enable secure communication between board members, fostering collaboration and timely decision-making.

Here are some practical examples of board management software that can enhance governance sustainability in your organization:

- **Diligent Boards:** Diligent Boards is a widely-used board management software that helps streamline board communications, manage meeting materials, and enhance collaboration among board members. Features include secure document storage, real-time updates, annotation tools, and meeting scheduling capabilities.

- **BoardEffect:** BoardEffect is a cloud-based board management solution that offers a centralized platform for organizing board materials, managing meetings, and communicating securely. With customizable modules, it enables organizations to design workflows that align with their specific governance processes.

- **OnBoard:** OnBoard is a comprehensive board management platform that simplifies board meeting preparation, facilitates secure communication, and provides easy access to important documents. The platform offers features such as an e-signature tool, a meeting scheduler, and offline access to board materials.

- **Boardvantage:** Boardvantage is a board portal that streamlines board communications, document management, and meeting organization. The platform offers advanced security features, such as two-factor authentication, data encryption, and remote data wiping, ensuring the protection of sensitive governance information.

- **Boardable:** Boardable is a board management software designed for nonprofits and mission-driven organizations. It offers tools for meeting scheduling, agenda building, document storage, and secure communication. The platform also includes a polling feature, enabling board members to vote on key decisions remotely.

Compliance and risk management tools

Implement compliance and risk management software to automate regulatory compliance monitoring, risk assessments, and reporting. By automating these processes, you can more efficiently identify potential compliance gaps, assess risks, and prioritize mitigation efforts.

Here are some practical examples of compliance and risk management tools that can enhance governance sustainability in your organization:

- **NAVEX Global:** NAVEX Global offers a suite of integrated risk and compliance management software solutions, including policy and procedure management, risk assessment, incident reporting, and third-party risk management. These tools help organizations streamline their compliance processes, identify risks, and implement effective controls.

- **LogicGate Risk Cloud:** LogicGate Risk Cloud is a customizable platform that enables organizations to automate their risk and compliance management processes. It offers modules for risk assessment, policy management, incident management, and compliance reporting, helping organizations stay on top of their regulatory obligations and mitigate risks.
- **RSA Archer:** RSA Archer is an integrated risk management platform that helps organizations manage multiple dimensions of risk, including operational, regulatory, and strategic risks. The platform offers a range of modules, such as audit management, policy management, and risk assessment, enabling organizations to identify, assess, and mitigate risks effectively.
- **MetricStream:** MetricStream provides a suite of governance, risk, and compliance (GRC) management solutions that help organizations automate and streamline their risk and compliance processes. Its platform includes tools for policy management, risk assessment, compliance monitoring, and regulatory change management.
- **Galvanize HighBond:** Galvanize HighBond is a GRC platform that offers a range of tools for risk management, compliance, and audit management. Its features include risk assessments, control testing, policy management, and automated compliance monitoring, helping organizations identify and address potential compliance gaps and risks.

Data analytics and visualization

Leverage data analytics and visualization tools to gain insights into your organization's performance, risk exposure, and governance effectiveness. These tools can help you identify trends, correlations, and anomalies, supporting more informed decision-making and strategic planning.

Here are some practical examples of data analytics and visualization tools that can enhance governance sustainability in your organization:

- **Tableau:** Tableau is a powerful data analytics and visualization tool that helps organizations analyze, understand, and visualize their ESG data related to governance sustainability. By using Tableau, companies can monitor trends in executive compensation, assess the impact of board diversity on performance, and identify areas of potential risk related to regulatory compliance.
- **Power BI:** Power BI is a business intelligence tool that allows organizations to analyze and visualize complex ESG data sets, including governance-related metrics. With Power BI, you can create interactive dashboards and reports to share insights with your board and stakeholders, helping to inform decision-making and drive governance improvements.
- **Qlik Sense:** Qlik Sense is a data analytics and visualization platform that enables organizations to explore and analyze governance sustainability data from multiple sources. With its drag-and-drop interface, Qlik Sense allows you to create custom visualizations and dashboards that highlight key governance metrics, trends, and insights, facilitating better communication with stakeholders and supporting informed decision-making.

E-voting and shareholder engagement platforms

Utilize e-voting and shareholder engagement platforms to facilitate remote voting and communication with shareholders. These platforms can increase shareholder participation in decision-making and help you gather valuable feedback on governance practices.

Here are some practical examples of e-voting and shareholder engagement platforms that can enhance governance sustainability in your organization:

- **Broadridge:** Broadridge is a leading provider of technology solutions for shareholder engagement and proxy voting. Its e-voting platform enables shareholders to cast their votes electronically, streamlining the proxy voting process and reducing costs for companies. Broadridge also offers a suite of shareholder communication tools, such as online forums and survey tools, that help companies engage with their shareholders and enhance transparency.

- **Mediant:** Mediant is another leading provider of shareholder engagement and e-voting solutions. Its platform allows shareholders to vote electronically, view their voting history, and access governance-related documents and information. Mediant also offers a range of communication tools, such as virtual meeting platforms and investor relations websites, that help companies engage with their shareholders and enhance governance sustainability.

- **Lumi AGM:** Lumi AGM is a web-based platform that allows companies to conduct virtual and hybrid annual general meetings (AGMs) and other shareholder meetings. With features such as live streaming, Q&A sessions, and electronic voting, Lumi AGM provides a seamless and secure way for shareholders to participate in meetings remotely. The platform also offers reporting and analytics tools that help companies monitor and measure shareholder engagement.

- **Shareworks:** Shareworks is a platform that helps companies manage their equity plans and engage with shareholders. Its shareholder engagement tools include an online shareholder portal, which allows shareholders to access their account information, vote electronically, and view governance-related documents. Shareworks also offers reporting and analytics features, such as vote results analysis and shareholder engagement metrics, that help companies assess the effectiveness of their governance sustainability efforts.

Artificial intelligence (AI) and machine learning

Apply AI and machine learning algorithms to analyze large datasets, detect anomalies, and identify potential risks or compliance issues. This technology can help you proactively address governance challenges and improve overall performance.

Here are some practical examples of artificial intelligence (AI) and machine learning tools that can enhance governance sustainability in your organization:

- **AI-powered risk assessment:** AI and machine learning algorithms can be used to analyze large volumes of data from various sources, helping to identify potential governance risks and vulnerabilities. These tools can analyze patterns and trends in areas such as regulatory compliance, board performance, and executive compensation, allowing you to prioritize risk mitigation efforts and improve overall governance.

- **Natural language processing (NLP) for regulatory compliance:** NLP is a subfield of AI that focuses on enabling computers to understand and interpret human language. By using NLP tools, organizations can automate the analysis of regulatory documents, policy updates, and news articles, ensuring that they stay up-to-date with relevant laws and regulations, and identify potential compliance issues.

- **AI-driven ethical decision-making:** AI and machine learning tools can be used to support ethical decision-making within your organization. These tools can analyze data related to business ethics, anti-corruption efforts, and other governance aspects, helping you to identify potential ethical risks and make informed decisions that align with your organization's values and principles.

- **AI-enhanced board evaluation:** AI-powered tools can help organizations conduct more efficient and effective board evaluations. By analyzing data from board meeting minutes, performance metrics, and stakeholder feedback, AI algorithms can provide insights into board dynamics, effectiveness, and areas for improvement, ultimately enhancing governance sustainability.

Cybersecurity tools

Strengthen your organization's cybersecurity posture by implementing advanced security tools and technologies, such as intrusion detection systems, encryption, and multi-factor authentication. Robust cybersecurity measures are essential to protect sensitive governance information and maintain stakeholder trust.

Incorporating these technologies into your governance practices will allow you to enhance the effectiveness and sustainability of your organization's governance efforts, ensuring that they remain agile, transparent, and responsive to evolving business needs and stakeholder expectations.

Here are some practical examples of cybersecurity tools that can enhance governance sustainability in your organization:

- **Endpoint protection:** Endpoint protection software helps safeguard your organization's devices, such as laptops, desktops, and mobile phones, from potential cyber threats. By protecting sensitive governance data and ensuring secure communication among board members and stakeholders, endpoint protection tools contribute to good governance practices.

- **Secure file sharing and storage:** Secure file sharing and storage platforms, such as Box or Microsoft SharePoint, ensure that sensitive governance-related documents and information are stored and shared securely. These tools often provide encryption, access control, and version tracking features, which help maintain the integrity and confidentiality of governance data.

- **Intrusion detection and prevention systems (IDPS):** IDPS tools monitor your organization's network for potential cyber threats and unauthorized access. By detecting and preventing intrusions, these tools help protect sensitive governance data and contribute to overall governance sustainability.

- **Security information and event management (SIEM) systems:** SIEM systems collect and analyze security data from various sources within your organization, helping to identify potential security risks and incidents. By providing real-time monitoring, analysis, and reporting on security events, SIEM tools support governance sustainability by ensuring the protection of critical governance data.

- **Two-factor authentication (2FA):** Implementing 2FA adds an extra layer of security to the authentication process, reducing the risk of unauthorized access to sensitive governance information. By requiring users to provide additional verification, such as a temporary code sent to their mobile device, 2FA helps maintain the integrity and confidentiality of governance data.

Module 4 Exam Questions

1. Which of the following is a key aspect of environmental sustainability?

a) Climate change and greenhouse gas emissions

b) Employee health, safety, and well-being

c) Board composition and diversity

d) Shareholder rights and engagement

2. Which component of ESG focuses on human rights and labor standards?

a) Environmental sustainability

b) Social sustainability

c) Governance sustainability

d) None of the above

3. What is a primary goal of corporate governance structures and policies?

a) Ensuring efficient resource management

b) Maintaining high levels of biodiversity

c) Promoting accountability and transparency

d) Reducing pollution and waste

4. Which of these is an example of technology for governance sustainability?

a) Solar panels

b) Board management software

c) Recycling systems

d) Electric vehicles

5. What is the purpose of ESG reporting frameworks and standards?

a) To outline environmental regulations

b) To guide companies in disclosing their ESG performance

c) To enforce labor laws

d) To set industry-specific sustainability goals

6. Which of the following is an example of a green bond?

a) A bond issued to finance environmentally friendly projects

b) A bond issued to fund community engagement programs

c) A bond issued to support executive compensation

d) A bond issued for general corporate purposes

7. What is the main focus of pollution prevention and control?

a) Developing eco-friendly products

b) Encouraging employee diversity

c) Reducing waste generation and emissions

d) Engaging with local communities

8. Which of the following is a key aspect of social sustainability?

a) Sustainable procurement and supply chain management

b) Environmental management systems

c) Diversity, equity, and inclusion

d) Climate change and greenhouse gas emissions

9. What is the primary goal of shareholder rights and engagement?

a) To protect the environment

b) To promote fair labor practices

c) To empower shareholders in corporate decision-making

d) To comply with regulatory requirements

10. Which of the following is a practical KPI for board composition and diversity?

a) Percentage of board members with relevant industry experience

b) Percentage of women and minorities on the board

c) Number of environmental certifications achieved

d) Employee satisfaction rates

11. What is the primary purpose of business ethics and anti-corruption measures?

a) To prevent financial losses

b) To protect the company's reputation

c) To comply with regulatory requirements

d) All of the above

12. Which of the following is an example of a compliance and risk management tool?

a) NAVEX Global

b) OnBoard

c) Boardable

d) Boardvantage

13. What is a key goal of talent attraction, retention, and development?

a) Ensuring efficient resource management

b) Building a skilled and diverse workforce

c) Reducing greenhouse gas emissions

d) Engaging with local communities

14. What is the primary objective of environmental management systems?

a) To monitor and improve an organization's environmental performance

b) To promote employee well-being

c) To engage with shareholders

d) To comply with labor laws

15. Which of the following is an example of a board management software?

a) LogicGate Risk Cloud

b) RSA Archer

c) MetricStream

d) Boardable

16. Which of the following is a component of governance sustainability?

a) Biodiversity and ecosystem preservation

b) Human rights and labor standards

c) Executive compensation and performance management

d) Product responsibility and customer safety

17. What is the purpose of ESG integration in portfolio management?

a) To reduce portfolio risk and enhance long-term returns

b) To support labor unions

c) To lobby for stricter environmental regulations

d) To improve employee satisfaction

18. Which of the following is an aspect of community engagement and development?

a) Reducing waste generation

b) Supporting local economic growth and social initiatives

c) Ensuring board diversity

d) Implementing corporate governance policies

19. What is the main focus of product responsibility and customer safety?

a) Ensuring products are safe and meet customer expectations

b) Promoting equal opportunities for all employees

c) Establishing a diverse board of directors

d) Engaging with shareholders on corporate decisions

20. Which of the following is a key aspect of regulatory compliance and risk management?

a) Minimizing the organization's environmental impact

b) Promoting employee health, safety, and well-being

c) Ensuring adherence to relevant laws and regulations

d) Encouraging diversity, equity, and inclusion

Answers:

1. a
2. b
3. c
4. b
5. b
6. a
7. c
8. c
9. c
10. b
11. d
12. a
13. b
14. a
15. d
16. c
17. a
18. b
19. a
20. c

Module 5. ESG Reporting and Disclosure

Introduction to ESG Reporting and Disclosure

Welcome to Module 5 on ESG Reporting and Disclosure. In this module, you will learn about the critical role that transparent and accurate reporting plays in the broader context of sustainability. A comprehensive and well-structured ESG reporting framework helps businesses communicate their sustainability performance, progress, and future goals to various stakeholders, including investors, employees, customers, and regulators.

Through practical examples, you will gain insights into the importance of ESG reporting in building trust and credibility with stakeholders, as well as in attracting and retaining responsible investments. This module will also highlight how effective ESG reporting can help companies uncover risks, identify opportunities, and drive continuous improvement in their sustainability performance.

As part of this module, you will explore how leading companies are leveraging best practices in ESG reporting to demonstrate their commitment to sustainability and to enhance their reputation in the market. You will also gain an understanding of how ESG reporting can be used as a tool to engage and involve stakeholders in a company's sustainability journey, creating a strong foundation for long-term success and value creation.

By the end of this module, you will be better equipped to appreciate the significance of robust ESG reporting and disclosure practices, and their impact on a company's overall ESG performance, competitiveness, and resilience in the face of emerging global challenges.

ESG Reporting Frameworks and Standards

In this section, you will explore various ESG reporting frameworks and standards that can help guide your organization in developing and presenting a comprehensive and transparent sustainability report. These frameworks and standards serve as essential tools to ensure consistency, comparability, and reliability in your ESG disclosures, enabling you to effectively communicate your sustainability performance to stakeholders.

Some of the widely recognized ESG reporting frameworks and standards include:

- **Global Reporting Initiative (GRI):** GRI is a leading, independent organization that develops global sustainability reporting guidelines. The GRI Standards provide a common language for organizations to report on their economic, environmental, and social performance, ensuring your sustainability report addresses critical issues in a structured and coherent manner.

  Example: If your company is in the oil and gas sector, you can use the GRI Oil & Gas standard (GRI 11) to report on your energy use, efficiency initiatives, and renewable energy sources.

- **Sustainability Accounting Standards Board (SASB):** SASB offers industry-specific ESG reporting standards that help you identify and disclose material sustainability issues relevant to your sector. By adopting SASB standards, you can ensure your ESG disclosures are both financially material and decision-useful for investors.

  Example: If you operate in the healthcare sector, SASB standards will guide you in disclosing topics such as patient privacy, pharmaceutical waste management, and access to medicines.

- **Task Force on Climate-related Financial Disclosures (TCFD):** TCFD provides a framework for companies to disclose information on their climate-related risks and opportunities, allowing you to demonstrate your organization's preparedness for the challenges and opportunities arising from climate change.

  Example: If your business operates in a region prone to extreme weather events, you can use TCFD recommendations to report on your climate resilience strategies, such as investing in infrastructure improvements or developing business continuity plans.

- **United Nations Global Compact (UNGC):** As a voluntary initiative, UNGC encourages organizations to align their strategies and operations with ten universal principles related to human rights, labor, environment, and anti-corruption. By participating in the UNGC, you can demonstrate your commitment to responsible business practices and contribute to the achievement of the United Nations Sustainable Development Goals (SDGs).

  Example: You can use UNGC's Communication on Progress (COP) template to report on your company's actions to promote gender equality, reduce water pollution, or implement ethical sourcing policies.

- **Carbon Disclosure Project (CDP):** CDP is a global non-profit organization that runs a disclosure system for companies, cities, and regions to report their environmental impacts, including greenhouse gas emissions, water usage, and deforestation. By participating in the CDP, you can showcase your organization's commitment to transparency and accountability in addressing climate change, water security, and deforestation.

Example: You can use CDP's reporting framework to disclose your company's carbon emissions reduction targets, water management practices, or efforts to combat deforestation in your supply chain.

- **S&P Global Corporate Sustainability Assessment (CSA):** The CSA is an annual evaluation conducted by S&P Global that assesses companies' sustainability performance across a range of ESG criteria. By participating in the CSA, you can benchmark your organization's ESG performance against industry peers, identify areas for improvement, and enhance your company's visibility among investors who use ESG ratings to inform their investment decisions.

  Example: If your company is involved in the consumer goods industry, the CSA will assess your sustainability performance across various aspects, such as product safety, labor practices, and environmental management. By addressing these criteria in your ESG reporting, you can demonstrate your commitment to sustainability and potentially improve your ESG ratings in the eyes of investors and other stakeholders.

As you explore these ESG reporting frameworks and standards, it's essential to consider the unique characteristics and objectives of your organization. By selecting the most suitable framework or adopting a combination of standards, you can create a robust and tailored ESG report that effectively communicates your sustainability performance, progress, and ambitions to your stakeholders. This, in turn, will help build trust, enhance your reputation, and contribute to the long-term success of your organization.

Materiality Assessment and Stakeholder Engagement

In this section, you will learn about the importance of conducting a materiality assessment and engaging with stakeholders to ensure that your ESG reporting is relevant, focused, and responsive to their needs and expectations.

Materiality Assessment

A materiality assessment helps you identify and prioritize the ESG topics that are most relevant to your organization and stakeholders. By conducting a materiality assessment, you can ensure that your ESG reporting focuses on the issues that matter most, both from a business risk and opportunity perspective and in terms of stakeholder concerns.

To perform a materiality assessment, you should:

1. **Identify relevant ESG topics:** Begin by creating a list of potential ESG topics that could be relevant to your organization, considering factors such as industry-specific issues, regulatory requirements, and sustainability trends.

   Example: If you operate in the manufacturing sector, your list of ESG topics might include energy efficiency, waste management, worker health and safety, and supply chain responsibility.

2. **Engage stakeholders:** Gather input from your key stakeholders, such as investors, customers, employees, suppliers, and regulators, to understand their perspectives on the importance of various ESG topics. This can be done through surveys, interviews, workshops, or other engagement methods.

   Example: You might conduct a survey among your investors to gauge their level of concern about your organization's carbon emissions, water usage, and labor practices.

3. **Prioritize topics:** Based on the feedback from stakeholders and your own internal analysis of business risks and opportunities, rank the ESG topics in terms of their importance. This will help you focus your reporting efforts on the most material issues.

   Example: After analyzing stakeholder feedback and your organization's risks and opportunities, you might determine that waste management, energy efficiency, and worker health and safety are the top three material topics for your ESG report.

Stakeholder Engagement

Engaging with stakeholders throughout the ESG reporting process is crucial for building trust, fostering collaboration, and ensuring that your disclosures are aligned with their expectations. Stakeholder engagement helps you validate the materiality of your ESG topics and collect valuable insights that can inform your sustainability strategy and reporting.

To effectively engage stakeholders in your ESG reporting process, you should:

1. **Identify key stakeholders:** Determine which stakeholder groups are most relevant to your organization's sustainability performance and have a vested interest in your ESG disclosures.

   Example: Key stakeholders for your ESG reporting might include investors, customers, employees, suppliers, regulators, and local communities.

2. **Establish a dialogue:** Develop a process for ongoing communication with stakeholders, providing them with opportunities to share their feedback, concerns, and expectations related to your ESG performance.

Example: You might organize regular stakeholder roundtables or webinars to discuss your organization's sustainability progress, challenges, and future plans.

3. **Integrate stakeholder input:** Use the insights gained from stakeholder engagement to inform and enhance your ESG reporting, ensuring that your disclosures address their concerns and priorities.

   Example: If your stakeholders express concerns about your organization's supply chain practices, you might expand your reporting to include more information on your supplier code of conduct, auditing processes, and supplier capacity building initiatives.

Conducting a robust materiality assessment and actively engaging with your stakeholders will allow you to create a focused, relevant, and responsive ESG report that effectively communicates your organization's sustainability performance and demonstrates your commitment to addressing the concerns of your stakeholders.

The Double Materiality Assessment

In this section, you will learn about the concept of double materiality and how it can help you enhance the quality and relevance of your ESG reporting by considering both the impact of ESG factors on your organization and your organization's impact on society and the environment.

Double materiality is an approach that encompasses two distinct dimensions of materiality:

- **Financial materiality:** This aspect focuses on the ESG factors that have a significant impact on your organization's financial performance, such as risks and opportunities related to climate change, resource scarcity, or regulatory compliance. By identifying and reporting on these financially material factors, you can provide decision-useful information to investors and demonstrate your organization's commitment to managing ESG-related risks and opportunities.

  Example: If your organization operates in a water-intensive industry, such as beverages or textiles, water scarcity could pose a significant financial risk. In this case, you should assess and report on your water management strategies, including water efficiency measures, alternative water sources, and collaboration with local stakeholders to address shared water challenges.

- **Societal and environmental materiality:** This dimension focuses on the ESG factors that reflect your organization's impact on society and the environment, regardless of their financial materiality. By identifying and reporting on these can demonstrate your organization's commitment to responsible business practices, contribute to the achievement of global sustainability goals, and enhance your reputation among stakeholders.

Example: If your organization operates in the technology sector, you might not face significant financial risks from your waste generation. However, your electronic waste could have severe environmental and social consequences, such as pollution and negative health impacts on local communities. In this case, you should assess and report on your strategies for reducing waste, recycling materials, and managing the disposal of electronic waste responsibly.

To conduct a double materiality assessment, you should follow these steps:

1. Identify relevant ESG factors: Start by creating a comprehensive list of potential ESG factors, considering both financial materiality (risks and opportunities for your organization) and societal and environmental materiality (your organization's impact on society and the environment).

2. Assess financial materiality: Analyze the potential financial risks and opportunities associated with each ESG factor, considering factors such as regulatory requirements, industry trends, and investor expectations.

3. Assess societal and environmental materiality: Evaluate the significance of each ESG factor in terms of its impact on society and the environment, considering factors such as stakeholder concerns, sustainable development goals, and industry best practices.

4. Prioritize ESG factors: Based on your analysis, prioritize the ESG factors according to their financial materiality and societal and environmental materiality. This will help you focus your reporting efforts on the most relevant and impactful factors.

5. Engage stakeholders: Validate your double materiality assessment by engaging with key stakeholders, such as

investors, customers, employees, and regulators. Their input can help you refine your understanding of the materiality of ESG factors and ensure that your reporting is aligned with stakeholder expectations.

Adopting a double materiality approach in your ESG reporting will allow you to provide a more comprehensive and balanced view of your organization's sustainability performance, addressing both the financial risks and opportunities associated with ESG factors and your organization's impact on society and the environment. This, in turn, will enhance the credibility, relevance, and value of your ESG disclosures for your stakeholders.

Data Collection and Quality Management

In this section, you will learn about the importance of effective data collection and quality management processes for your ESG reporting. Accurate, reliable, and consistent data is essential for creating a credible and informative ESG report that effectively communicates your sustainability performance to your stakeholders.

To ensure the integrity of your ESG data, you should consider the following best practices for data collection and quality management:

- **Develop a data collection plan:** Establish a clear and structured process for collecting ESG data from various sources within your organization, such as operational systems, surveys, and external databases. Your plan should include a list of data points to be collected, responsible parties, data sources, and timelines.

  Example: You might create a data collection template for each department in your organization, outlining the specific ESG metrics they need to report, the data sources to be used, and the deadlines for data submission.

- **Standardize data collection methodologies:** Adopt consistent methodologies for measuring and reporting on your ESG metrics, ensuring that your data is comparable across different business units, locations, and time periods.

  Example: You can use the Greenhouse Gas (GHG) Protocol to calculate and report your organization's carbon emissions in a standardized manner, facilitating comparisons with industry peers and tracking your progress over time.

- **Implement data validation and verification processes:** Establish procedures for checking the accuracy, completeness, and consistency of your ESG data, such as internal reviews, cross-referencing with other data sources, and third-party verification.

  Example: You might require that each department's ESG data undergoes a review by a designated data quality manager, who checks the data for accuracy and consistency before it is included in your ESG report.

- **Provide training and support for data collectors:** Ensure that the individuals responsible for collecting and reporting ESG data within your organization have the necessary knowledge, skills, and resources to perform their tasks effectively.

  Example: You could offer training workshops for your data collectors, covering topics such as ESG reporting frameworks, data collection methodologies, and data quality management best practices.

- **Maintain comprehensive documentation:** Keep detailed records of your ESG data collection and quality management processes, including data sources, methodologies, assumptions, and any changes made over time. This documentation can serve as an essential reference for internal and external stakeholders seeking to understand and assess the credibility of your ESG reporting.

  Example: You might create an ESG data inventory, detailing the data points collected, the measurement methodologies used, and the data sources and assumptions applied, ensuring transparency and traceability in your reporting process.

Implementing these best practices for data collection and quality management will allow you to ensure that your ESG report is underpinned by accurate, reliable, and consistent data, enhancing the credibility and value of your sustainability disclosures for your stakeholders. This, in turn, will help build trust, foster accountability, and support informed decision-making among your investors, customers, employees, and other stakeholders.

ESG Communication and Transparency

In this section, you will learn about the importance of effective communication and transparency in your ESG reporting. Clear, accessible, and transparent communication is crucial for building trust, fostering stakeholder engagement, and demonstrating your organization's commitment to sustainability.

To enhance your ESG communication and transparency, consider the following best practices:

- **Use clear and concise language:** Ensure that your ESG report is written in a straightforward and understandable manner, avoiding technical jargon and complex terms that may be difficult for your stakeholders to grasp.

  Example: Instead of using industry-specific acronyms, provide clear explanations and definitions to help your stakeholders better understand your sustainability performance and initiatives.

- **Provide context and benchmarks:** Help your stakeholders interpret your ESG data by providing relevant context and benchmarks, such as industry averages, historical trends, and comparisons with peers.

  Example: When reporting your organization's energy consumption, you can include a comparison with industry benchmarks and a historical analysis to demonstrate your progress towards greater energy efficiency.

- **Be transparent about challenges and limitations:** Openly acknowledge the challenges, limitations, and uncertainties associated with your ESG data and initiatives, providing an honest and balanced view of your sustainability performance.

Example: If your organization faced challenges in meeting its carbon reduction targets, you can explain the factors that contributed to this outcome and the steps you are taking to address these challenges moving forward.

- **Highlight your ESG strategy and goals:** Clearly communicate your organization's ESG strategy, objectives, and targets, demonstrating your commitment to continuous improvement and long-term sustainability.

  Example: In your ESG report, you can include a dedicated section outlining your sustainability vision, goals, and strategic initiatives, helping stakeholders understand how your organization is addressing ESG risks and opportunities.

- **Engage with stakeholders through multiple channels:** Use various communication channels to share your ESG report and engage with your stakeholders, such as your website, social media, webinars, and in-person events.

  Example: You can publish your ESG report on your corporate website, promote its release through social media, and host webinars or roundtable discussions to engage with your stakeholders and gather their feedback on your sustainability performance.

- **Seek external assurance:** Enhance the credibility of your ESG report by obtaining external assurance from a reputable third-party provider, such as an auditing firm or a sustainability consultancy.

  Example: You can engage an external assurance provider to review your ESG data, methodologies, and reporting processes, and include their assurance statement in your report to demonstrate your commitment to transparency and accuracy.

Adopting these best practices for ESG communication and transparency will allow you to effectively convey your organization's sustainability performance, challenges, and ambitions to your stakeholders, building trust and fostering a meaningful dialogue around your ESG initiatives. This, in turn, can contribute to improved stakeholder relations, enhanced reputation, and long-term business success.

Technology for ESG Reporting and Disclosure

In this section, you will learn about the role of technology in facilitating and enhancing ESG reporting and disclosure. By leveraging technology tools and platforms, you can streamline your data collection, analysis, and reporting processes, ensuring that your ESG disclosures are accurate, reliable, and timely.

To effectively utilize technology for your ESG reporting and disclosure, consider the following tools and approaches:

- **Data management systems:** Implement a centralized data management system to collect, store, and manage your ESG data, ensuring consistency, accuracy, and accessibility. Such systems can help you automate data collection, aggregation, and validation processes, saving time and reducing the risk of errors.

  Example: You can use an ESG data management platform that integrates with your existing operational systems, such as your enterprise resource planning (ERP) system, to automatically collect and process ESG data from various sources within your organization.

- **Analytics and visualization tools:** Leverage analytics and visualization tools to analyze and present your ESG data in an engaging and meaningful way. These tools can help you identify trends, patterns, and relationships in your data, facilitating data-driven decision-making and stakeholder engagement.

  Example: You can use a business intelligence (BI) platform to create interactive dashboards and visualizations that showcase your organization's ESG performance, enabling your stakeholders to explore and interpret your ESG data more easily.

- **ESG reporting software:** Utilize ESG reporting software to streamline the process of creating and publishing your ESG reports. Such software can help you comply with various reporting frameworks and standards, generate pre-defined templates, and automate the production of charts, tables, and graphs.

  Example: You can use an ESG reporting tool that supports the Global Reporting Initiative (GRI) Standards, the Sustainability Accounting Standards Board (SASB) Framework, and other relevant reporting guidelines, ensuring that your ESG report aligns with best practices and stakeholder expectations.

- **Artificial intelligence (AI) and machine learning:** Harness the power of AI and machine learning to enhance your ESG data analysis, identify emerging risks and opportunities, and develop predictive models for your sustainability performance.

  Example: You can use machine learning algorithms to analyze large volumes of ESG data, such as news articles, social media posts, and regulatory filings, to identify emerging sustainability trends and assess their potential impact on your organization.

- **Blockchain and distributed ledger technology:** Explore the potential of blockchain and distributed ledger technology to enhance the transparency, traceability, and security of your ESG data, particularly in areas such as supply chain management, carbon credits, and impact measurement.

  Example: You can use blockchain technology to track and verify the environmental and social performance of your suppliers, creating a transparent and tamper-proof record of your supply chain sustainability efforts.

Leveraging these technology tools and approaches will allow you to enhance the efficiency, accuracy, and impact of your ESG reporting and disclosure, providing your stakeholders with a more comprehensive, accessible, and meaningful understanding of your organization's sustainability performance. This, in turn, can contribute to improved decision-making, stakeholder relations, and long-term business success.

Practical ESG reporting software tools

Let's discuss some practical ESG reporting software tools that can help you streamline your ESG data management, analysis, and reporting processes. These tools can enhance the efficiency, accuracy, and impact of your ESG reporting and disclosure, making it easier for you to communicate your organization's sustainability performance to your stakeholders.

- **Workiva:** Workiva is a cloud-based platform that simplifies ESG reporting by automating data collection, aggregation, and validation processes. It offers pre-built templates for various reporting frameworks, such as GRI, SASB, and TCFD, and allows for seamless collaboration among team members. Workiva also provides data visualization tools to help you create engaging and interactive ESG reports.

- **Enablon:** Enablon is an integrated ESG and risk management software that supports data collection, analysis, and reporting for a wide range of sustainability metrics. It offers customizable dashboards, automated reporting capabilities, and compliance with various ESG frameworks and standards. Enablon also enables you to track and manage your sustainability initiatives, measure their impact, and identify areas for improvement.

- **SAP Control Tower:** SAP Control Tower is a supply chain-focused ESG reporting tool that provides end-to-end visibility of your supply chain sustainability performance. It integrates with your existing SAP systems and offers real-time data tracking, predictive analytics, and risk management capabilities. SAP Control Tower can help you monitor your suppliers' environmental and social performance, track the progress of your sustainability initiatives, and ensure compliance with relevant regulations and standards.

- **Datamaran:** Datamaran is an AI-powered ESG risk management and reporting platform that uses machine learning to analyze large volumes of data from various sources, such as regulatory filings, news articles, and social media. It helps you identify emerging ESG trends, assess their potential impact on your organization, and align your reporting with best practices and stakeholder expectations. Datamaran also provides benchmarking tools, enabling you to compare your ESG performance with industry peers and track your progress over time.

- **Emex EHS & ESG Software:** Emex is a comprehensive EHS (Environment, Health, and Safety) and ESG management platform that enables you to collect, analyze, and report on a wide range of sustainability metrics. It offers automated data collection and validation, customizable dashboards, and compliance with various ESG frameworks and standards. Emex also provides tools for managing your EHS and ESG initiatives, tracking their impact, and identifying areas for improvement.

- **Envizi by IBM:** Envizi is a powerful ESG data management and reporting platform owned by IBM. It enables organizations to collect, analyze, and report on a wide range of environmental, social, and governance metrics. Envizi offers a user-friendly interface, customizable dashboards, and automated data collection and validation processes. It supports various ESG reporting frameworks and standards, such as GRI, SASB, and TCFD, ensuring that your disclosures align with best practices and stakeholder expectations. In addition, Envizi integrates with IBM's suite of sustainability solutions, allowing you to leverage advanced analytics, AI, and blockchain technologies to enhance your ESG performance and reporting capabilities.

Leveraging these ESG reporting software tools will allow you to streamline your data collection, analysis, and reporting processes, ensuring that your ESG disclosures are accurate, reliable, and timely. This can help you effectively convey your organization's sustainability performance to your stakeholders, build trust, and foster meaningful engagement around your ESG initiatives.

Module 5 Exam Questions

1. Which of the following is a key component of Sustainability?

a) Environmental Sustainability

b) Governance Sustainability

c) Social Sustainability

d) All of the above

2. Which of the following ESG reporting frameworks focuses on industry-specific metrics?

a) Global Reporting Initiative (GRI)

b) Sustainability Accounting Standards Board (SASB)

c) Task Force on Climate-related Financial Disclosures (TCFD)

d) Carbon Disclosure Project (CDP)

3. What is a primary focus of environmental sustainability?

a) Climate change and greenhouse gas emissions

b) Employee health, safety, and well-being

c) Board composition and diversity

d) Human rights and labor standards

4. Which aspect of social sustainability addresses workplace diversity and equal opportunities?

a) Human rights and labor standards

b) Employee health, safety, and well-being

c) Diversity, equity, and inclusion

d) Community engagement and development

5. What is a key aspect of governance sustainability?

a) Corporate governance structures and policies

b) Resource management and waste reduction

c) Talent attraction, retention, and development

d) Biodiversity and ecosystem preservation

6. Which ESG reporting framework focuses on climate-related risks and opportunities?

a) Global Reporting Initiative (GRI)

b) Sustainability Accounting Standards Board (SASB)

c) Task Force on Climate-related Financial Disclosures (TCFD)

d) Carbon Disclosure Project (CDP)

7. What is the purpose of a materiality assessment in ESG reporting?

a) Identifying the most relevant ESG issues for your organization

b) Analyzing your organization's carbon footprint

c) Benchmarking your organization against industry peers

d) Assessing your organization's financial performance

8. Which of the following is a practical ESG reporting software tool?

a) Workiva

b) Enablon

c) SAP Control Tower

d) All of the above

9. What technology can be used to enhance the transparency and traceability of ESG data?

a) Artificial intelligence

b) Blockchain and distributed ledger technology

c) Data visualization tools

d) Machine learning algorithms

10. Which ESG reporting standard is focused on carbon emissions and climate change?

a) Global Reporting Initiative (GRI)

b) Sustainability Accounting Standards Board (SASB)

c) Task Force on Climate-related Financial Disclosures (TCFD)

d) Carbon Disclosure Project (CDP)

11. What is the primary goal of sustainable procurement and supply chain management?

a) Reducing waste and conserving resources

b) Ensuring ethical and environmentally responsible sourcing

c) Promoting energy efficiency

d) Fostering innovation in product design

12. Which aspect of governance sustainability addresses executive compensation and performance management?

a) Board composition and diversity

b) Corporate governance structures and policies

c) Executive compensation and performance management

d) Shareholder rights and engagement

13. What is the main objective of ESG integration in portfolio management?

a) Minimizing financial risks associated with ESG factors

b) Maximizing short-term financial returns

c) Focusing solely on environmental factors

d) Ignoring social and governance factors

14. Which ESG reporting software tool is developed by IBM?

a) Workiva

b) Enablon

c) SAP Control Tower

d) Envizi

15. In the context of ESG reporting, what does the double materiality assessment consider?

a) Financial and non-financial risks

b) Environmental and social risks

c) Short-term and long-term risks

d) Risks and opportunities

16. Which aspect of social sustainability focuses on product responsibility and customer safety?

a) Human rights and labor standards

b) Employee health, safety, and well-being

c) Diversity, equity, and inclusion

d) Product responsibility and customer safety

17. In the context of ESG, what is the main goal of regulatory compliance and risk management?

a) Ensuring adherence to laws, regulations, and industry standards

b) Reducing the organization's environmental footprint

c) Promoting workplace diversity and inclusion

d) Engaging with local communities and stakeholders

18. What is the primary purpose of green bonds and sustainable finance instruments?

a) Raising capital for environmentally-friendly projects

b) Divesting from fossil fuel companies

c) Increasing short-term financial returns

d) Reducing the organization's overall debt

19. Which ESG reporting framework is operated by a non-profit organization that requests companies to disclose their environmental data?

a) Global Reporting Initiative (GRI)

b) Sustainability Accounting Standards Board (SASB)

c) Task Force on Climate-related Financial Disclosures (TCFD)

d) Carbon Disclosure Project (CDP)

20. What role does technology play in the future of Sustainability?

a) Enhancing data collection and analysis

b) Facilitating better communication and transparency

c) Fostering innovation and new solutions

d) All of the above

Answers:

1. d
2. b
3. a
4. c
5. a
6. c
7. a
8. d
9. b
10. d
11. b
12. c
13. a
14. d
15. d
16. d
17. a
18. a
19. d
20. d

Module 6. ESG Integration and Strategy

Introduction to ESG Integration and Strategy

Welcome to Module 6 on ESG Integration and Strategy. In this module, you will learn the importance of incorporating ESG factors into your organization's strategic planning and decision-making processes. We will discuss how businesses can effectively integrate ESG principles into their core operations, thereby driving sustainable growth and long-term value creation.

To achieve successful ESG integration, organizations need to move beyond simply treating ESG as a separate, standalone initiative. Instead, they must embed ESG considerations into all aspects of their business, from risk management and product development to supply chain management and stakeholder engagement. By doing so, organizations can better anticipate and respond to ESG-related risks and opportunities, ultimately enhancing their resilience, competitiveness, and reputation.

Throughout this module, we will provide practical examples of how leading companies have successfully integrated ESG into their strategies and operations. These real-world case studies will illustrate how organizations can leverage ESG insights to identify new markets, develop innovative products and services, and improve their overall performance. We will also explore the critical role that leadership plays in driving ESG integration and fostering a culture of sustainability within an organization.

By the end of this module, you will have a deeper understanding of the importance of ESG integration and the practical steps that your organization can take to embed ESG principles into its strategy and operations. You will also be equipped with the knowledge and tools needed to create a comprehensive ESG roadmap, allowing you to turn ESG insights into actionable steps and drive positive change within your organization.

ESG Risk and Opportunity Assessment

Conducting an ESG risk and opportunity assessment is an essential step in the integration of ESG principles into your business strategy. By identifying and evaluating the potential ESG risks and opportunities that your company faces, you can make informed decisions and allocate resources effectively. This process will help you understand how ESG factors can impact your organization's financial performance, reputation, and long-term value creation.

To begin the ESG risk and opportunity assessment process, you should follow these steps:

1. **Identify ESG factors relevant to your business:** Start by considering the environmental, social, and governance factors that may affect your company's operations, industry, and geographical location. For example, climate change might pose risks related to regulatory compliance, physical impacts, and supply chain disruptions.

2. **Evaluate the significance of each factor:** Assess the potential impact of each identified ESG factor on your business, taking into account the likelihood and severity of the consequences. This will help you prioritize your efforts and focus on the most significant ESG risks and opportunities.

3. **Map ESG factors to your business activities:** Determine how each ESG factor relates to your company's core business activities, such as product development, procurement, and manufacturing. This will enable you to integrate ESG considerations into your operational processes more effectively.

4. **Assess your current ESG performance:** Evaluate your company's current ESG performance by benchmarking against industry standards, peers, and best practices. This will help you identify gaps, strengths, and areas for improvement.

5. **Identify potential ESG opportunities:** Recognize potential opportunities for your company to create value by addressing ESG issues. These opportunities might include developing sustainable products and services, improving operational efficiency, or tapping into new markets.

For example, if your company operates in the manufacturing sector, you might identify energy efficiency as a key ESG factor. You can assess the significance of this factor by considering the potential cost savings, regulatory requirements, and reputational benefits associated with improving energy efficiency. By mapping this factor to your production processes, you can identify specific opportunities to reduce energy consumption, such as upgrading equipment or optimizing production schedules.

Once you have conducted a comprehensive ESG risk and opportunity assessment, you can use the insights gained to inform your business strategy and decision-making processes. By addressing the most significant ESG risks and capitalizing on the identified opportunities, you will be better positioned to create long-term value for your company and its stakeholders.

Developing an ESG Strategy and Roadmap

Once you have conducted an ESG risk and opportunity assessment, the next step is to develop an ESG strategy and roadmap. This process will help you set clear goals, prioritize actions, and establish a plan to integrate ESG factors into your business operations effectively. Here's how you can develop an ESG strategy and roadmap:

1. **Set clear ESG objectives:** Based on your ESG risk and opportunity assessment, define clear and measurable objectives that align with your company's overall mission, values, and strategic goals. These objectives should address the most significant ESG risks and capitalize on the identified opportunities. For example, you may aim to reduce your greenhouse gas emissions by 20% within five years or achieve a specific diversity target in your leadership team.

2. **Establish priorities and targets:** Prioritize the ESG factors that are most relevant to your business and set specific targets for each. Ensure that these targets are ambitious, yet achievable, and have clear deadlines. For instance, you might prioritize waste reduction and set a target to decrease waste generation by 30% within three years.

3. **Develop an action plan:** Create a detailed action plan outlining the steps required to achieve your ESG objectives and targets. This plan should include specific initiatives, resource allocations, and timelines. For example, if your goal is to improve water efficiency, your action plan might involve implementing water-saving technologies, enhancing employee training, and setting up a water management system.

4. **Assign responsibilities and accountability:** Assign clear roles and responsibilities to your team members to ensure that everyone understands their part in implementing the ESG strategy. Establish accountability mechanisms, such as performance evaluations and regular progress reports, to monitor the execution of the plan and track progress toward your ESG goals.

5. **Engage stakeholders:** Involve relevant stakeholders, such as employees, investors, suppliers, and customers, in the development and implementation of your ESG strategy. This will help you gain valuable insights, build trust, and ensure that your ESG efforts address the needs and expectations of your stakeholders.

6. **Integrate ESG into decision-making:** Embed ESG considerations into your company's decision-making processes, including budgeting, capital allocation, and strategic planning. This will ensure that ESG factors are taken into account when making business decisions and help drive a culture of sustainability throughout your organization.

For example, suppose you run a clothing company and have identified sustainable sourcing as a priority. In that case, your ESG strategy might involve setting targets to increase the use of ethically sourced materials, working closely with suppliers to ensure compliance with social and environmental standards, and educating consumers about the importance of sustainable fashion. Your roadmap would then outline the specific actions, timelines, and resources required to achieve these objectives, as well as the mechanisms for tracking progress and ensuring accountability.

Developing a comprehensive ESG strategy and roadmap will allow you to effectively integrate ESG principles into your business operations and create long-term value for your company and its stakeholders.

ESG Performance Measurement and Monitoring

Measuring and monitoring your ESG performance is critical for ensuring the success of your ESG strategy and continuous improvement. Effective performance measurement and monitoring allow you to track progress, make data-driven decisions, and demonstrate your commitment to ESG to stakeholders. Here are the steps to measure and monitor your ESG performance effectively:

1. **Select relevant Key Performance Indicators (KPIs):** Choose specific, measurable KPIs that align with your ESG objectives and targets. These KPIs should enable you to track your progress towards your goals and evaluate the effectiveness of your ESG initiatives. For example, if your objective is to reduce greenhouse gas emissions, you might select a KPI such as "tons of CO_2 emissions per unit of production."

2. **Establish a baseline:** Determine your current performance level for each KPI to establish a baseline. This will allow you to measure your progress over time and assess the impact of your ESG initiatives. For instance, you might calculate your current water usage per unit of production to establish a baseline for your water efficiency KPI.

3. **Set performance targets:** Set ambitious but achievable performance targets for each KPI. These targets should be aligned with your overall ESG objectives and should have specific timeframes for achievement. For example, you might set a target to reduce your CO_2 emissions per unit of production by 25% within five years.

4. **Collect and analyze data:** Implement a system for collecting and analyzing data related to your ESG KPIs. This system should ensure data accuracy, consistency, and comparability over time. Regularly analyze the collected data to identify trends, progress, and areas for improvement. For instance, you might track your energy consumption data monthly to assess the effectiveness of your energy efficiency initiatives.

5. **Monitor progress and adjust your strategy:** Regularly monitor your progress towards your ESG targets and evaluate the effectiveness of your initiatives. If you find that your performance is not on track to achieve your targets, adjust your strategy and action plan accordingly. This might involve reallocating resources, refining your initiatives, or setting new targets.

6. **Communicate your ESG performance:** Share your ESG performance with stakeholders, such as investors, customers, and employees, through regular reporting and communication. This will help demonstrate your commitment to ESG and build trust with stakeholders. For example, you might publish an annual sustainability report that highlights your ESG achievements, challenges, and future plans.

A practical example of ESG performance measurement and monitoring might be a company in the food and beverage industry aiming to improve its packaging sustainability. The company could establish KPIs such as "percentage of recyclable packaging materials" and "amount of packaging waste generated per product." By collecting data on these KPIs, setting performance targets, and monitoring progress, the company can assess the effectiveness of its packaging sustainability initiatives and make data-driven decisions to improve its ESG performance.

Implementing a robust ESG performance measurement and monitoring system will allow you to ensure continuous improvement in your ESG initiatives and demonstrate your commitment to sustainability to your stakeholders.

Continuous Improvement and Innovation in ESG

Achieving excellence in ESG requires a commitment to continuous improvement and innovation. By constantly seeking ways to enhance your ESG performance, you can stay ahead of industry trends, regulatory changes, and stakeholder expectations. Here are some steps to foster continuous improvement and innovation in your ESG initiatives:

1. **Embrace a culture of learning:** Encourage a learning mindset within your organization by providing ongoing ESG training and education for your employees. This will help build their capacity to identify and address ESG risks and opportunities and drive continuous improvement in your ESG performance.

2. **Set challenging goals:** Establish ambitious ESG objectives and targets that push your organization to improve its performance continually. By setting challenging goals, you will inspire your team to think creatively and find innovative solutions to ESG issues.

3. **Monitor industry trends and best practices:** Stay informed of emerging ESG trends, best practices, and regulatory developments in your industry. This will enable you to adapt your ESG strategy and initiatives proactively and ensure that you remain at the forefront of sustainability.

4. **Encourage innovation:** Foster a culture of innovation by supporting and rewarding employees for developing new ideas and solutions related to ESG. Encourage cross-functional collaboration and provide resources, such as time and funding, for employees to experiment with new ESG initiatives.

5. **Regularly review and update your ESG strategy:** Periodically assess your ESG strategy to ensure that it remains aligned with your organization's overall goals and evolving stakeholder expectations. Update your strategy as needed to address emerging risks and capitalize on new opportunities.

6. **Measure and celebrate success:** Recognize and celebrate your ESG achievements to reinforce the importance of continuous improvement and motivate your team to continue striving for excellence. Share your successes with stakeholders to demonstrate your commitment to sustainability and build trust.

For example, suppose you operate in the renewable energy sector, and you aim to increase the efficiency of your solar panels continually. In that case, you could encourage your research and development team to explore cutting-edge technologies and collaborate with external partners, such as academic institutions or startups, to drive innovation in solar panel efficiency. By setting ambitious targets for efficiency improvements and providing resources for experimentation, you can foster a culture of continuous improvement and innovation within your organization.

Prioritizing continuous improvement and innovation in ESG will allow you to ensure that your organization remains a leader in sustainability, adapts to changing conditions, and continues to create long-term value for your stakeholders.

Technology for ESG Integration and Strategy

Leveraging technology can play a significant role in the successful integration and implementation of your ESG strategy. Technology can help you collect, analyze, and report ESG data more efficiently, facilitate communication and collaboration, and drive innovation in your ESG initiatives. Here are some examples of how technology can support your ESG efforts:

1. **Data collection and management:** Utilize software tools and platforms to collect, store, and manage ESG data from various sources, such as energy consumption meters, waste management systems, or employee surveys. Examples of ESG data management tools include Enablon, Emisoft, and IsoMetrix, which can help you automate data collection, ensure data quality, and maintain data consistency.

2. **Data analysis and visualization:** Use analytics tools and software to analyze your ESG data, identify trends, and uncover insights that can inform your decision-making processes. Tools such as Tableau, Power BI, and SAS can help you visualize and interpret complex ESG data, enabling you to identify areas for improvement, monitor progress, and make data-driven decisions.

3. **ESG reporting and disclosure:** Leverage technology to streamline and standardize your ESG reporting processes. ESG reporting platforms like Workiva or Datamaran can help you generate reports based on various ESG frameworks and standards, ensuring that your reporting is consistent, transparent, and aligned with stakeholder expectations.

4. **Collaboration and communication:** Utilize digital communication platforms and tools, such as Microsoft Teams, Slack, or Asana, to facilitate collaboration among your team members and stakeholders. These tools can help you share ESG-related information, coordinate efforts, and track progress towards your ESG goals.

5. **ESG performance monitoring:** Implement software solutions to track your ESG performance and measure progress towards your targets. Tools like Sustainalytics, Trucost, and MSCI ESG Manager can provide you with real-time ESG performance data, enabling you to monitor progress, adjust your strategy, and ensure accountability.

6. **Innovation and continuous improvement:** Leverage technology to drive innovation in your ESG initiatives. For example, you might use artificial intelligence (AI) and machine learning (ML) to optimize your energy consumption patterns, or deploy Internet of Things (IoT) devices to monitor and manage your resource use more efficiently.

For instance, if your organization aims to reduce its carbon footprint, you could implement IoT-enabled sensors to collect real-time data on your energy consumption and use AI-powered analytics tools to identify patterns and opportunities for improvement. By leveraging these technologies, you can optimize your energy use, reduce your carbon emissions, and enhance your overall ESG performance.

Harnessing the power of technology will allow you to streamline your ESG integration and strategy, making it more efficient, effective, and adaptable to evolving needs and expectations.

Module 6 Exam Questions

1. Which aspect of ESG focuses on a company's relationships with employees, suppliers, customers, and the communities where it operates?

a) Environmental

b) Social

c) Governance

d) None of the above

2. What is a primary goal of environmental sustainability?

a) Reducing greenhouse gas emissions

b) Increasing shareholder value

c) Enhancing corporate governance structures

d) Improving employee well-being

3. Which of the following is an aspect of social sustainability?

a) Pollution prevention and control

b) Executive compensation and performance management

c) Employee health, safety, and well-being

d) Regulatory compliance and risk management

4. What is the purpose of ESG reporting and disclosure?

a) To provide information about a company's ESG performance to stakeholders

b) To measure a company's financial performance

c) To identify potential merger and acquisition targets

d) To assess a company's market share

5. What is a key step in developing an ESG strategy and roadmap?

a) Identifying ESG risks and opportunities

b) Calculating the company's market share

c) Analyzing competitor's marketing strategies

d) Assessing the company's intellectual property

6. What tool can be used to visualize and interpret complex ESG data?

a) Microsoft Word

b) Tableau

c) Adobe Photoshop

d) Microsoft Excel

7. Which aspect of ESG focuses on board composition and diversity?

a) Environmental

b) Social

c) Governance

d) None of the above

8. What is an example of an environmental sustainability initiative?

a) Sustainable procurement and supply chain management

b) Talent attraction, retention, and development

c) Business ethics and anti-corruption

d) Shareholder rights and engagement

9. What is a primary goal of governance sustainability?

a) Enhancing corporate governance structures and policies

b) Reducing greenhouse gas emissions

c) Increasing shareholder value

d) Improving employee well-being

10. In which ESG aspect do you focus on human rights and labor standards?

a) Environmental

b) Social

c) Governance

d) None of the above

11. What is an example of a social sustainability initiative?

a) Climate change and greenhouse gas emissions

b) Resource management and waste reduction

c) Community engagement and development

d) Pollution prevention and control

12. What type of software tools can help collect, store, and manage ESG data?

a) Data management tools

b) Graphic design tools

c) Audio editing tools

d) Video editing tools

13. Which of the following is an ESG reporting platform?

a) Enablon

b) GRI

c) Tableau

d) Microsoft Teams

14. What is an example of a governance sustainability initiative?

a) Biodiversity and ecosystem preservation

b) Employee health, safety, and well-being

c) Corporate governance structures and policies

d) Sustainable procurement and supply chain management

15. Which of the following can be used to facilitate collaboration among team members and stakeholders in ESG initiatives?

a) Microsoft Teams

b) Microsoft Word

c) Microsoft Excel

d) Adobe Photoshop

16. Which of the following is a key performance indicator (KPI) related to environmental sustainability?

a) Tons of CO2 emissions per unit of production

b) Percentage of women on the board of directors

c) Number of employee training hours

d) Total shareholder return

17. What is the purpose of a materiality assessment in ESG reporting and disclosure?

a) To identify the most significant ESG issues for a company and its stakeholders

b) To measure a company's financial performance

c) To assess a company's market share

d) To analyze competitors' marketing strategies

18. What technology can be used to drive innovation in ESG initiatives related to optimizing energy consumption patterns?

a) Artificial intelligence (AI) and machine learning (ML)

b) Graphic design software

c) Audio editing software

d) Spreadsheet software

19. Which of the following is an aspect of ESG integration and strategy?

a) ESG risk and opportunity assessment

b) ESG data collection and management

c) ESG reporting and disclosure

d) ESG investment and finance

20. Which ESG aspect focuses on reducing a company's impact on the environment?

a) Environmental

b) Social

c) Governance

d) None of the above

Answers:

1. b
2. a
3. c
4. a
5. a
6. b
7. c
8. a
9. a
10. b
11. c
12. a
13. b
14. c
15. a
16. a
17. a
18. a
19. a
20. a

Module 7. ESG in Investment and Finance

Introduction to ESG in Investment and Finance

Welcome to Module 7 on ESG in Investment and Finance. In this module, you will learn about the growing significance of ESG factors in the world of investment and finance. We will delve into the ways in which responsible investing has evolved over the years, and how it is shaping the decision-making processes of various stakeholders in the financial industry.

ESG integration in investment and finance has become increasingly important as investors, lenders, and other financial institutions recognize the potential risks and opportunities associated with environmental, social, and governance factors. By incorporating ESG considerations into their investment strategies, these institutions aim to generate sustainable, long-term returns while also promoting positive change in the companies they invest in or lend to.

Throughout this module, we will explore the concept of responsible investing and its various forms, such as impact investing, socially responsible investing (SRI), and thematic investing. We will also discuss how ESG factors are incorporated into the evaluation of investment opportunities and financial products, such as green bonds and other sustainable finance instruments.

Additionally, we will shed light on the role of ESG ratings and benchmarking in the investment process, providing practical examples of how these assessments help investors make informed decisions. Furthermore, we will discuss the importance of ESG integration in portfolio management, demonstrating how it contributes to effective risk management and the identification of new investment opportunities.

By the end of this module, you will have a thorough understanding of the critical role that ESG factors play in investment and finance, and you will be equipped with the knowledge to make informed decisions in this evolving landscape. Through practical examples and real-world scenarios, this module aims to provide you with valuable insights into the world of responsible investing and sustainable finance.

Responsible Investing and ESG Factors

Responsible investing refers to the practice of incorporating environmental, social, and governance (ESG) factors into investment decision-making. As an investor, you can use these factors to assess the sustainability and ethical implications of your investments, as well as their potential risks and returns. In this section, we will explore the importance of ESG factors in responsible investing, along with some practical examples to help you better understand the concept.

When you engage in responsible investing, you aim to generate positive, sustainable outcomes for society and the environment, alongside achieving financial returns. By considering ESG factors, you can better evaluate the long-term resilience and performance of companies in your investment portfolio.

Here are some examples of ESG factors that you may want to consider when making investment decisions:

Environmental factors:

- Climate change risks and opportunities, such as a company's greenhouse gas emissions, energy efficiency, and adaptation strategies.
- Waste management and resource efficiency, including recycling practices and the reduction of water consumption.
- Pollution prevention, including the management of hazardous waste and the reduction of air, water, and soil contamination.

Social factors:

- Human rights and labor standards, such as fair wages, safe working conditions, and policies that prevent child labor and forced labor.

- Employee health, safety, and well-being, including initiatives to improve workplace safety and support employee mental health.

- Diversity, equity, and inclusion, encompassing gender balance, equal opportunities, and anti-discrimination policies within the workforce.

Governance factors:

- Corporate governance structures and policies, including board composition, director independence, and accountability mechanisms.

- Executive compensation and performance management, assessing alignment between executive pay and company performance.

- Business ethics and anti-corruption, including the establishment of codes of conduct, policies on bribery and corruption, and whistleblowing mechanisms.

Incorporating ESG factors into your investment decisions will allow you to identify companies that are well-positioned to navigate emerging risks and capitalize on new opportunities. For example, companies with strong environmental management systems may be better prepared for regulatory changes related to climate change and have a competitive advantage in transitioning to a low-carbon economy.

As a practical example, let's say you are considering investing in a technology company. By assessing its ESG performance, you may discover that the company has a diverse and inclusive workforce, strong data privacy policies, and transparent supply chain practices. These factors can contribute to the company's resilience and long-term value creation, making it a more attractive investment option.

In conclusion, responsible investing and ESG factors are crucial for making informed decisions that align with your financial goals and values. By considering these factors, you can identify sustainable and ethically responsible investments that contribute to long-term value creation and positive societal and environmental impacts.

Green Bonds and Sustainable Finance Instruments

As you consider incorporating sustainable investment practices into your portfolio, it's essential to familiarize yourself with green bonds and other sustainable finance instruments. These financial products provide an opportunity to invest in projects and companies that focus on positive environmental and social outcomes while offering competitive returns. In this section, we'll explore the different types of sustainable finance instruments and provide practical examples to illustrate their benefits.

Green Bonds

Green bonds are debt securities issued by companies, governments, or other entities to finance environmentally friendly projects. These projects can include renewable energy, energy efficiency improvements, pollution control, sustainable transportation, and climate change adaptation measures. By investing in green bonds, you support projects that contribute to a more sustainable future while receiving interest payments and the return of your principal investment upon the bond's maturity.

For example, imagine a city government issuing a green bond to finance the construction of a new solar power plant. As an investor, you could purchase this bond, directly supporting the development of clean energy infrastructure while earning interest on your investment.

Sustainable Bonds

Sustainable bonds combine the features of green bonds and social bonds, financing projects with both environmental and social benefits. These bonds can support initiatives such as affordable housing, access to clean water and sanitation, or sustainable agriculture.

An example of a sustainable bond could be one issued by a company to fund the construction of energy-efficient, affordable housing units in underserved communities. By investing in this bond, you help address both environmental and social challenges while receiving returns on your investment.

Social Bonds

Social bonds are similar to green bonds but focus specifically on projects that deliver positive social outcomes. These projects may include healthcare, education, social inclusion, or job creation initiatives that benefit disadvantaged or marginalized populations.

A practical example of a social bond could be one issued by a nonprofit organization to finance the expansion of a job training program for unemployed individuals in low-income areas. Investing in this bond allows you to support the social objective while generating financial returns.

Transition Bonds

Transition bonds are issued to finance projects that help carbon-intensive industries shift towards more sustainable operations. These bonds can support companies in sectors such as energy, transportation, and manufacturing as they invest in new technologies and practices to reduce their environmental impact.

For instance, a steel manufacturer might issue a transition bond to fund the development of a new production process that significantly lowers carbon emissions. By investing in this bond, you contribute to the company's transition towards a more sustainable future.

In summary, green bonds and other sustainable finance instruments provide you with opportunities to invest in projects and companies that align with your values while generating competitive financial returns. By exploring these options, you can create a more diversified and sustainable investment portfolio that contributes to positive environmental and social outcomes.

ESG Ratings and Benchmarking

In order to make informed investment decisions that incorporate ESG factors, it's essential for you to understand ESG ratings and benchmarking. These tools help you evaluate the ESG performance of companies, making it easier to compare and select investments based on their sustainability and ethical practices. In this section, we will discuss the importance of ESG ratings and benchmarking, and provide practical examples to guide your decision-making process.

ESG ratings are assessments of companies' environmental, social, and governance performance provided by specialized research firms, rating agencies, or index providers. These ratings assign scores to companies based on a wide range of ESG indicators, such as greenhouse gas emissions, labor practices, and corporate governance structures. By comparing ESG ratings, you can identify companies that excel in managing ESG risks and seizing opportunities related to sustainability and ethical practices.

When using ESG ratings, it's crucial to be aware of the methodologies employed by different providers, as they can vary significantly in terms of the indicators they consider and the weightings they assign to various factors. Some well-known ESG rating providers include MSCI, Sustainalytics, and FTSE Russell, among others.

Benchmarking is the process of comparing a company's ESG performance against its peers or industry standards. This can help you identify leaders and laggards in terms of ESG practices, providing valuable insights into the companies that may be better positioned to thrive in a rapidly changing world. ESG benchmarks can be based on ESG ratings or specific performance metrics, such as carbon emissions or gender diversity on boards.

Now, let's consider a practical example to illustrate the use of ESG ratings and benchmarking in your investment decision-making process. Imagine you are interested in investing in the consumer goods sector and want to select companies with strong ESG performance. You could start by reviewing ESG ratings from a reputable provider to identify companies with high scores in the sector. Additionally, you could benchmark these companies against industry averages or best practices to further refine your selection.

For instance, you might find that Company A has an ESG rating above the industry average, demonstrating strong performance in reducing its carbon footprint, promoting diversity and inclusion, and ensuring transparent governance. In contrast, Company B might have a below-average ESG rating due to its weak labor practices and lack of commitment to reducing its environmental impact. Based on this information, you may choose to invest in Company A, as it exhibits stronger ESG performance and appears to be better positioned for long-term success.

In conclusion, ESG ratings and benchmarking are valuable tools that can help you make more informed investment decisions based on companies' sustainability and ethical practices. By incorporating these assessments into your investment process, you can better align your portfolio with your values and support companies that demonstrate a commitment to ESG excellence.

ESG Integration in Portfolio Management

Integrating ESG factors into your portfolio management process is essential for achieving long-term, sustainable returns while aligning your investments with your values. ESG integration helps you effectively manage risks, identify opportunities, and enhance the overall performance of your investment portfolio. In this section, we will discuss various approaches to ESG integration and provide practical examples to help you effectively incorporate these factors into your portfolio management strategy.

Here are some key steps and approaches for integrating ESG factors into your portfolio management process:

1. **Establish your ESG objectives:** Start by defining your ESG goals and risk tolerance. This will help you determine the extent to which you want to integrate ESG factors into your investment strategy and the specific criteria you will use to evaluate and select investments.

2. **ESG research and analysis:** Conduct thorough research and analysis of potential investments, incorporating ESG data from rating agencies, research firms, or other sources. This will help you gain a comprehensive understanding of companies' ESG performance and identify the risks and opportunities associated with their environmental, social, and governance practices.

3. **Security selection:** Use your ESG research and analysis to inform your security selection process. You may choose to invest in companies with strong ESG ratings or those that demonstrate leadership in specific ESG areas, such as climate change mitigation or diversity and inclusion. You might also consider avoiding or divesting from companies with poor ESG performance or exposure to significant ESG risks.

4. **Portfolio construction and optimization:** As you construct and optimize your investment portfolio, consider the diversification benefits of incorporating a variety of ESG-focused investments across different sectors and asset classes. This can help you manage risks and enhance the resilience of your portfolio in the face of changing market conditions.

5. **Active ownership and engagement:** As an investor, you can influence the ESG practices of companies in your portfolio through active ownership and engagement. This may involve voting on shareholder resolutions related to ESG issues, engaging with company management to encourage improvements in ESG performance, or collaborating with other investors to amplify your impact.

6. **Monitoring and reporting:** Regularly monitor and evaluate the ESG performance of your investments and adjust your portfolio as needed to ensure alignment with your ESG objectives. Communicate your ESG integration approach and the resulting portfolio performance to stakeholders through transparent reporting.

For a practical example, let's assume you are managing a diversified equity portfolio and want to integrate ESG factors into your investment strategy. You could begin by establishing your ESG objectives, such as focusing on companies with strong climate change mitigation strategies or those with exemplary labor practices. Next, you could conduct research and analysis using ESG data to identify companies that meet your criteria.

Once you have selected your ESG-focused investments, you can construct and optimize your portfolio by incorporating these securities across various sectors and industries, ensuring proper diversification. As an active owner, you could engage with the management of the companies in your portfolio to encourage ongoing improvement in their ESG performance. Finally, monitor and report on the ESG performance of your portfolio to demonstrate your commitment to sustainable investing and keep your stakeholders informed.

In conclusion, ESG integration in portfolio management is an essential aspect of responsible investing, allowing you to align your investment strategy with your values and manage ESG-related risks and opportunities. By following the steps and approaches outlined in this section, you can build a resilient and sustainable investment portfolio that contributes to positive environmental, social, and governance outcomes.

Technology for ESG in Investment and Finance

Leveraging technology in the investment and finance sector can help you effectively integrate ESG factors into your decision-making process, improve data analysis, and enhance your overall investment strategy. In this section, we will discuss various technologies that support ESG integration in investment and finance, along with practical examples to demonstrate their applications.

- **Data analytics and artificial intelligence (AI):** Advanced data analytics tools and AI-driven algorithms can process large volumes of ESG data from various sources, such as corporate disclosures, news articles, and social media, to provide you with valuable insights into companies' ESG performance. For example, you can use natural language processing (NLP) algorithms to analyze companies' sustainability reports and identify key ESG risks and opportunities that may not be evident through traditional analysis.

- **Portfolio optimization software:** ESG-focused portfolio optimization software enables you to construct and manage your investment portfolio by incorporating ESG factors and constraints. These tools can help you analyze the ESG characteristics of your portfolio, identify areas for improvement, and simulate the impact of potential investment decisions on your ESG performance and risk profile.

- **Environmental, social, and governance (ESG) data platforms:** ESG data platforms aggregate ESG information from multiple sources, providing you with comprehensive and reliable data to inform your investment decisions. These platforms may offer ESG ratings, benchmarking tools, and industry-specific metrics, making it easier for you to evaluate and compare the ESG performance of companies in your portfolio or potential investments.

- **Blockchain and distributed ledger technology (DLT):** Blockchain and DLT can enhance transparency and traceability in the ESG space, enabling you to track the environmental and social impact of your investments more accurately. For example, blockchain-based supply chain tracking systems can help you verify the sustainability credentials of products and materials, ensuring that your investments support responsible sourcing practices.
- **Climate scenario analysis tools:** Climate scenario analysis tools allow you to assess the potential impact of various climate change scenarios on your investments. By simulating the effects of different climate policies, physical risks, and transition pathways, these tools can help you identify climate-related risks and opportunities and build a more resilient portfolio.

A practical example of technology in ESG investment and finance could involve using a combination of data analytics tools, AI algorithms, and an ESG data platform to analyze the ESG performance of companies in your investment universe. You can then employ portfolio optimization software to construct an ESG-focused portfolio that aligns with your objectives and risk tolerance. Additionally, you might utilize blockchain technology to verify the sustainability credentials of your investments and use climate scenario analysis tools to assess the potential impact of climate change on your portfolio.

In conclusion, technology plays a critical role in ESG integration for investment and finance, offering a range of tools and solutions to help you effectively manage ESG risks and opportunities. By leveraging these technologies, you can enhance your investment decision-making process, build a more sustainable and resilient portfolio, and contribute to positive environmental, social, and governance outcomes.

Software tools for ESG in Investment and Finance

Incorporating software tools designed specifically for ESG in investment and finance can help you effectively manage and analyze ESG data, make informed investment decisions, and optimize your portfolio. In this section, we will discuss various software tools that can support your ESG integration efforts and provide practical examples to demonstrate their applications.

- **ESG data platforms:** ESG data platforms aggregate ESG information from multiple sources, offering comprehensive and reliable data to inform your investment decisions. These platforms often provide ESG ratings, benchmarking tools, and industry-specific metrics, making it easier for you to evaluate and compare the ESG performance of companies in your portfolio or potential investments. Examples of ESG data platforms include MSCI ESG Manager, Sustainalytics, and Bloomberg ESG Data Hub.

- **Portfolio optimization software:** ESG-focused portfolio optimization software enables you to construct and manage your investment portfolio by incorporating ESG factors and constraints. These tools can help you analyze the ESG characteristics of your portfolio, identify areas for improvement, and simulate the impact of potential investment decisions on your ESG performance and risk profile. Examples of portfolio optimization software include BlackRock Aladdin, FactSet, and Morningstar Direct.

- **ESG risk and opportunity assessment tools:** These software tools help you identify and quantify ESG risks and opportunities associated with your investments. By providing in-depth analysis of specific ESG factors, such as climate change, human rights, or corporate governance, these tools can enhance your understanding of ESG-related risks and inform your investment decisions.

Examples of ESG risk and opportunity assessment tools include Truvalue Labs, Datamaran, and RepRisk.

- **Climate scenario analysis tools:** Climate scenario analysis tools allow you to assess the potential impact of various climate change scenarios on your investments. By simulating the effects of different climate policies, physical risks, and transition pathways, these tools can help you identify climate-related risks and opportunities and build a more resilient portfolio. Examples of climate scenario analysis tools include Carbon Delta, 2° Investing Initiative, and CDP Scenario Analysis.

Let's consider a practical example of how you can use these software tools in your ESG investment and finance activities. Imagine you are managing an equity portfolio and want to integrate ESG factors into your investment strategy. First, you could utilize an ESG data platform, such as MSCI ESG Manager, to access comprehensive ESG data on companies in your investment universe. Next, you could employ portfolio optimization software like BlackRock Aladdin to incorporate ESG factors and construct a diversified portfolio that aligns with your objectives and risk tolerance.

Once your portfolio is established, you can use ESG risk and opportunity assessment tools, such as Truvalue Labs, to continuously monitor your investments for emerging ESG risks and opportunities. Additionally, you might employ climate scenario analysis tools like Carbon Delta to evaluate the potential impact of climate change on your portfolio and identify areas for improvement.

In conclusion, software tools play an essential role in ESG integration for investment and finance, offering a range of solutions to help you effectively manage ESG data, make informed investment decisions, and optimize your portfolio. By leveraging these tools, you can enhance your investment decision-making process, build a more sustainable and resilient portfolio, and contribute to positive environmental, social, and governance outcomes.

Module 7 Exam Questions

1. Which of the following is a component of environmental sustainability?

a) Human rights and labor standards

b) Climate change and greenhouse gas emissions

c) Board composition and diversity

d) Product responsibility and customer safety

2. What is the primary goal of sustainable procurement and supply chain management?

a) Reducing carbon emissions

b) Ensuring ethical labor practices

c) Minimizing negative environmental and social impacts

d) Maximizing profits

3. Which aspect of social sustainability focuses on fair treatment and equal opportunities for employees?

a) Diversity, equity, and inclusion

b) Employee health, safety, and well-being

c) Human rights and labor standards

d) Community engagement and development

4. What is the purpose of ESG reporting frameworks and standards?

a) To promote corporate transparency and accountability

b) To eliminate all ESG risks

c) To rank companies based on ESG performance

d) To ensure that all companies adopt the same ESG practices

5. Which of the following is a key step in developing an ESG strategy and roadmap?

a) ESG risk and opportunity assessment

b) Materiality assessment and stakeholder engagement

c) Regulatory compliance and risk management

d) ESG communication and transparency

6. What is the main objective of responsible investing?

a) Maximizing short-term returns

b) Aligning investments with ethical values and long-term sustainability

c) Investing only in green technology companies

d) Avoiding investments in companies with poor governance

7. Green bonds are an example of which type of financial instrument?

a) Conventional bonds

b) Sustainable finance instruments

c) Cryptocurrency

d) Derivatives

8. What is the primary purpose of ESG ratings and benchmarking?

a) To compare companies based on their ESG performance

b) To create a standardized set of ESG metrics

c) To rank companies by market capitalization

d) To assign a financial value to a company's ESG practices

9. Which technology can improve transparency and traceability in the ESG space?

a) Artificial intelligence

b) Virtual reality

c) Blockchain and distributed ledger technology

d) Quantum computing

10. In portfolio management, what is the purpose of active ownership and engagement?

a) To increase the market value of the portfolio

b) To influence the ESG practices of companies in the portfolio

c) To minimize taxes on investment gains

d) To eliminate all ESG risks from the portfolio

11. What is an example of an ESG data platform?

a) MSCI ESG Manager

b) Microsoft Excel

c) Adobe Photoshop

d) Google Analytics

12. What is a key benefit of using portfolio optimization software for ESG integration?

a) Enhancing portfolio diversification

b) Eliminating all ESG risks

c) Maximizing short-term returns

d) Ensuring that all investments are green

13. Which ESG risk and opportunity assessment tool focuses on analyzing corporate disclosures, news articles, and social media?

a) Carbon Delta

b) Truvalue Labs

c) BlackRock Aladdin

d) Morningstar Direct

14. What is the primary goal of climate scenario analysis tools?

a) To predict future stock prices

b) To assess the potential impact of climate change on investments

c) To optimize a portfolio's asset allocation

d) To eliminate all climate-related risks from a portfolio

15. Which of the following is a component of governance sustainability?

a) Resource management and waste reduction

b) Employee health, safety, and well-being

c) Corporate governance structures and policies

d) Sustainable procurement and supply chain management

16. What is the main objective of ESG integration in portfolio management?

a) Maximizing short-term returns

b) Minimizing taxes on investment gains

c) Aligning investments with ESG factors and long-term sustainability

d) Investing only in green technology companies

17. Which aspect of environmental sustainability focuses on the efficient use of resources and minimizing waste?

a) Resource management and waste reduction

b) Climate change and greenhouse gas emissions

c) Biodiversity and ecosystem preservation

d) Pollution prevention and control

18. What is a primary goal of community engagement and development in social sustainability?

a) Maximizing company profits

b) Reducing greenhouse gas emissions

c) Fostering positive relationships and creating shared value with local communities

d) Ensuring all employees receive the same compensation

19. In the context of ESG investment and finance, what is the purpose of using data analytics and artificial intelligence (AI) tools?

a) To replace human decision-making

b) To process large volumes of ESG data and provide valuable insights

c) To predict future ESG trends with 100% accuracy

d) To eliminate the need for ESG data platforms

20. Which of the following is a benefit of using blockchain and distributed ledger technology (DLT) in ESG investment and finance?

a) Eliminating the need for ESG risk assessments

b) Increasing portfolio diversification

c) Enhancing transparency and traceability in ESG-related activities

d) Reducing the importance of corporate governance structures and policies

Answers:

1. b
2. c
3. a
4. a
5. a
6. b
7. b
8. a
9. c
10. b
11. a
12. a
13. b
14. b
15. c
16. c
17. a
18. c
19. b
20. c

Module 8. Case Studies and Best Practices in ESG Sustainability

Introduction to Case Studies and Best Practices in ESG Sustainability

Welcome to Module 8 on Case Studies and Best Practices in ESG Sustainability. In this module, you will learn about the practical implementation of ESG strategies and initiatives across various industries, as well as the value of cross-sector collaboration and partnerships in driving sustainability outcomes. We will dive deep into real-life examples, illustrating how organizations have successfully integrated ESG principles into their core business operations and achieved meaningful progress toward their sustainability goals.

Throughout this module, we will highlight the innovative approaches taken by leading companies in different sectors to address complex ESG challenges, and how these actions have positively impacted their financial performance, brand reputation, and stakeholder relationships. By examining these case studies, you will gain a better understanding of how organizations can navigate the complexities of ESG sustainability and identify best practices that can be adapted to your own organization's unique context.

In addition to exploring industry-specific initiatives and innovations, we will also discuss the importance of cross-sector collaboration and partnerships in scaling up sustainable solutions and amplifying their impact. By learning from the experiences of others and leveraging the power of collective action, your organization can accelerate its transition toward a more sustainable future.

By the end of this module, you will have a deeper understanding of the practical aspects of ESG implementation, enabling you to apply the insights and lessons learned from these case studies to your own organization's ESG journey. Armed with this knowledge, you will be better equipped to overcome obstacles, seize opportunities, and drive meaningful change in the pursuit of a more sustainable world.

Industry-Specific ESG Initiatives and Innovations

As you explore the world of ESG sustainability, it's essential to understand that industry-specific initiatives and innovations play a crucial role in addressing unique challenges and opportunities within each sector. In this section, we will discuss various examples of industry-specific ESG initiatives, providing you with practical insights and ideas that you can incorporate into your own sustainability efforts.

Energy Sector

The energy sector has been at the forefront of the sustainability movement, with a significant focus on transitioning to renewable energy sources and reducing greenhouse gas emissions. For example, you could consider investing in solar, wind, or hydroelectric power generation to decrease your dependence on fossil fuels. Additionally, you might explore energy efficiency initiatives, such as upgrading to LED lighting or implementing energy management systems, to reduce overall energy consumption and costs.

Automotive Industry

In the automotive industry, the shift towards electric vehicles (EVs) has been a transformative force. As a player in this sector, you could prioritize the development and promotion of electric or hybrid vehicles, invest in charging infrastructure, and collaborate with battery manufacturers to improve battery technology and recycling. Moreover, you can focus on enhancing fuel efficiency and reducing emissions for conventional vehicles, while also working on sustainable manufacturing processes and materials.

Textile and Apparel Industry

Sustainable fashion has become increasingly important in the textile and apparel industry. You can implement initiatives such as sourcing sustainable materials, reducing water and energy consumption during production, and adopting circular economy principles, such as recycling and upcycling. Furthermore, you can work on establishing ethical labor practices and fair wages throughout your supply chain to promote social sustainability.

Food and Beverage Industry

In the food and beverage industry, sustainable agriculture and responsible sourcing have gained prominence. You can promote regenerative farming practices, support local and organic producers, and adopt measures to reduce food waste throughout your supply chain. Additionally, consider investing in plant-based or alternative protein sources to address the environmental and ethical concerns associated with animal agriculture.

Finance Sector

ESG integration in the finance sector is rapidly evolving, with responsible investing and sustainable finance instruments gaining traction. As a financial institution, you can develop and offer green bonds, sustainable loans, and ESG-focused investment products, as well as integrate ESG factors into your risk assessment and decision-making processes. This approach will enable you to align your portfolio with long-term sustainability goals and better serve clients with sustainability preferences.

Telecommunications Sector

The telecommunications sector plays a critical role in the global transition to a more sustainable future by enabling digital connectivity and innovative technologies. As a player in this industry, you can prioritize energy efficiency in your operations, invest in renewable energy sources, and reduce electronic waste by implementing recycling and take-back programs for devices.

Additionally, you can work on expanding access to digital services in underserved communities, promoting digital literacy, and ensuring data privacy and security for your customers.

Construction and Real Estate Sector

Sustainable construction and real estate practices can have a significant impact on reducing the environmental footprint of buildings and urban environments. As a stakeholder in this sector, you can adopt green building standards, such as LEED or BREEAM, to design and construct energy-efficient, water-saving, and environmentally friendly buildings. You can also consider incorporating green spaces, promoting sustainable urban planning, and retrofitting existing buildings to improve their environmental performance. Social sustainability can be addressed by ensuring affordable housing options and prioritizing the well-being of local communities.

Healthcare Sector

The healthcare sector has unique opportunities to contribute to ESG sustainability by focusing on responsible resource management, waste reduction, and patient-centered care. You can implement waste reduction and recycling programs, prioritize energy-efficient equipment, and work on reducing the environmental impact of pharmaceuticals and medical supplies. Additionally, you can promote diversity, equity, and inclusion in the workplace, ensure access to affordable and quality healthcare services, and support community health initiatives.

Tourism and Hospitality Sector

Sustainable tourism and hospitality practices can help minimize the environmental and social impacts of travel while promoting responsible consumption and local development. As a business in this industry, you can adopt eco-friendly practices, such as reducing water and energy consumption, implementing waste management programs, and sourcing locally produced and sustainable products. Furthermore, you can support the local economy by promoting local artisans, businesses, and cultural experiences and engage with local communities to ensure that tourism development benefits all stakeholders.

Retail Sector

In the retail sector, there are numerous opportunities to integrate ESG principles into your operations and supply chain. You can work on reducing packaging waste, promoting circular economy principles through product take-back and recycling programs, and sourcing products that align with fair labor and environmental standards. Additionally, you can implement energy efficiency measures in your facilities and transportation, engage with local communities, and prioritize diversity and inclusion within your workforce.

Agriculture Sector

The agriculture sector plays a vital role in addressing global food security and environmental challenges. As a stakeholder in this industry, you can implement sustainable farming practices, such as precision agriculture, integrated pest management, and agroforestry, to enhance productivity while minimizing environmental impacts. Additionally, you can promote soil health, water conservation, and biodiversity protection and work to reduce greenhouse gas emissions associated with agricultural activities. Social sustainability can be addressed by ensuring fair labor practices, promoting rural development, and supporting smallholder farmers.

Technology Sector

In the technology sector, ESG initiatives can help drive innovation and improve the sustainability performance of products and services. As a player in this industry, you can focus on reducing the environmental footprint of your products by designing energy-efficient devices, using sustainable materials, and implementing responsible end-of-life management, such as recycling and e-waste reduction programs. You can also address social sustainability by promoting digital inclusion, data privacy, and security, as well as fostering a diverse and inclusive workplace culture.

Pharmaceutical and Biotechnology Sector

The pharmaceutical and biotechnology sectors can contribute to ESG sustainability by focusing on research and development for innovative, accessible, and affordable healthcare solutions. As a stakeholder in this industry, you can prioritize responsible supply chain management, reduce the environmental impact of manufacturing processes, and ensure proper disposal of medical waste. Additionally, you can address social sustainability by promoting access to essential medicines, supporting public health initiatives, and fostering a diverse, inclusive, and skilled workforce.

Transportation and Logistics Sector

The transportation and logistics sector plays a critical role in reducing greenhouse gas emissions and promoting sustainable supply chain management. As a business in this industry, you can invest in energy-efficient and low-emission transportation modes, such as electric or hydrogen-powered vehicles and rail transport. You can also implement route optimization, vehicle maintenance, and driver training programs to minimize fuel consumption and emissions. Social sustainability can be addressed by ensuring fair labor practices, workplace safety, and community engagement.

Consumer Goods Sector

In the consumer goods sector, companies can integrate ESG principles throughout their operations, from product design to end-of-life management. You can focus on developing products with sustainable materials and eco-friendly packaging, reducing waste and promoting circular economy principles. Additionally, you can work to ensure ethical and responsible sourcing practices throughout your supply chain, addressing issues such as deforestation, human rights, and labor standards. You can also prioritize diversity, equity, and inclusion in your workforce and engage with local communities to create shared value.

Education Sector

The education sector has a unique opportunity to contribute to ESG sustainability by shaping the next generation of leaders, innovators, and change-makers. As an institution in this sector, you can integrate sustainability principles into your curriculum and research activities, fostering an interdisciplinary understanding of global challenges and solutions. You can also promote campus sustainability through energy and water conservation, waste reduction, and sustainable procurement practices. Social sustainability can be addressed by ensuring access to quality education for all, promoting diversity and inclusion, and engaging with local and global communities.

Mining and Metals Sector

The mining and metals sector can contribute to ESG sustainability by addressing environmental and social challenges associated with resource extraction and processing. As a stakeholder in this industry, you can prioritize responsible mining practices, such as efficient resource use, land rehabilitation, and minimizing water pollution. Additionally, you can invest in clean technologies to reduce greenhouse gas emissions and energy consumption. Social sustainability can be addressed by ensuring fair labor practices, health and safety standards, and engaging with local communities to minimize the negative impacts of mining activities and create shared benefits.

Chemicals Sector

In the chemicals sector, ESG initiatives can focus on reducing the environmental impact of chemical production, waste management, and ensuring product safety. As a company in this industry, you can invest in cleaner production processes, energy-efficient technologies, and responsible waste disposal practices. You can also prioritize the development of environmentally friendly and biodegradable chemical products. Social sustainability can be addressed by ensuring workplace safety, fair labor practices, and engaging with stakeholders to promote product stewardship and responsible chemical management.

Media and Entertainment Sector

The media and entertainment sector has the power to shape public opinion and drive positive change through responsible content creation and sustainable operations. As a player in this industry, you can focus on promoting diversity, equity, and inclusion in content production and representation, as well as addressing issues such as digital privacy, security, and misinformation. You can also work on reducing the environmental footprint of your operations through energy efficiency, waste reduction, and sustainable procurement practices. Social sustainability can be addressed by fostering a diverse and inclusive workplace culture and supporting community development initiatives.

Water and Wastewater Management Sector

In the water and wastewater management sector, ESG initiatives can focus on responsible resource management, infrastructure development, and reducing pollution. As a stakeholder in this industry, you can prioritize water efficiency and conservation practices, invest in resilient and sustainable infrastructure, and implement advanced treatment technologies to minimize the release of pollutants into the environment. Social sustainability can be addressed by ensuring access to clean water and sanitation for all, promoting water stewardship, and engaging with local communities to address their water-related needs and concerns.

Aerospace and Defense Sector

The aerospace and defense sector can contribute to ESG sustainability by focusing on reducing the environmental impact of its products and operations, as well as addressing social and ethical concerns. As a company in this industry, you can invest in fuel-efficient technologies, alternative propulsion systems, and lightweight materials to minimize the environmental footprint of aircraft and other products. You can also prioritize responsible supply chain management, fair labor practices, and ethical conduct in business dealings. Social sustainability can be addressed by fostering a diverse and inclusive workforce, promoting employee safety and well-being, and engaging with stakeholders to address the social and environmental implications of your products and services.

Professional Services Sector

In the professional services sector, ESG initiatives can focus on responsible business practices, sustainable operations, and contributing to the broader sustainability agenda through client engagements. As a company in this industry, you can prioritize diversity, equity, and inclusion in your workforce, ensure fair labor practices, and promote employee well-being. You can also work on reducing the environmental footprint of your offices and operations through energy efficiency, waste reduction, and sustainable procurement practices. By advising clients on ESG integration and sustainability strategies, you can contribute to positive change across various industries and sectors.

Non-profit and NGO Sector

Non-profits and NGOs play a critical role in driving ESG sustainability by addressing environmental, social, and governance issues and advocating for responsible policies and practices. As an organization in this sector, you can focus on specific ESG issues or adopt a more holistic approach to sustainability. You can implement sustainable operations by prioritizing energy efficiency, waste reduction, and responsible resource management. Social sustainability can be addressed by promoting diversity and inclusion, fostering a healthy work culture, and ensuring transparent governance structures.

Maritime and Shipping Sector

The maritime and shipping sector can contribute to ESG sustainability by focusing on reducing the environmental impact of shipping operations, promoting responsible supply chain management, and addressing social concerns. As a company in this industry, you can invest in fuel-efficient technologies, alternative fuels, and cleaner propulsion systems to minimize greenhouse gas emissions and other pollutants. You can also prioritize responsible vessel management, waste reduction, and ballast water treatment. Social sustainability can be addressed by ensuring fair labor practices, promoting crew welfare, and working to combat illegal activities such as human trafficking and illegal fishing.

Waste Management Sector

In the waste management sector, ESG initiatives can focus on promoting circular economy principles, reducing waste generation, and minimizing environmental impacts. As a stakeholder in this industry, you can invest in advanced waste treatment technologies, such as waste-to-energy, recycling, and composting facilities, to reduce landfill disposal and greenhouse gas emissions. You can also promote waste reduction and recycling programs, engage with local communities to raise awareness, and collaborate with businesses to implement sustainable waste management practices. Social sustainability can be addressed by ensuring fair labor practices, health and safety standards, and transparent governance structures.

Forestry and Pulp and Paper Sector

The forestry and pulp and paper sector can contribute to ESG sustainability by focusing on responsible forest management, resource efficiency, and reducing the environmental impact of production processes. As a company in this industry, you can prioritize sustainable forestry practices, such as certification schemes like FSC or PEFC, to ensure biodiversity conservation and carbon sequestration. You can also invest in clean technologies, energy efficiency, and water conservation to minimize the environmental footprint of pulp and paper production. Social sustainability can be addressed by ensuring fair labor practices, community engagement, and respecting the rights and interests of indigenous peoples.

Examining these diverse sectors will allow you to gain a broader understanding of the range of ESG initiatives and innovations that can be tailored to your industry's specific context. Learning from these examples will help you identify best practices and drive meaningful progress toward a more sustainable future.

Cross-Sector Collaboration and Partnerships

Cross-sector collaboration and partnerships are essential for driving ESG sustainability, as they enable stakeholders from different industries and sectors to pool their resources, expertise, and influence to address complex and interconnected sustainability challenges. By engaging in cross-sector collaboration, you can identify innovative solutions, scale up your impact, and accelerate progress towards a more sustainable future. Here are some practical examples of how cross-sector collaboration and partnerships can work:

Public-Private Partnerships (PPPs)

Public-private partnerships involve collaboration between government entities and private sector organizations to develop, finance, and implement projects that address pressing sustainability challenges. By participating in PPPs, you can leverage public resources and expertise, access new markets, and share risks and responsibilities. For example, you can join forces with local governments to invest in renewable energy projects, develop sustainable infrastructure, or promote access to clean water and sanitation in underserved communities.

Collaborative Research and Development

Engaging in collaborative research and development initiatives with academia, research institutions, and other organizations can help you access cutting-edge knowledge, foster innovation, and develop new products, services, or technologies that address ESG challenges. For instance, you can collaborate with universities and research centers to explore new materials, alternative energy sources, or circular economy solutions that enhance the sustainability performance of your industry.

Industry Associations and Multi-Stakeholder Initiatives

Participating in industry associations or multi-stakeholder initiatives can provide you with a platform to collaborate with peers, competitors, and other stakeholders to address shared ESG challenges and develop best practices. For example, you can join initiatives such as the Global Reporting Initiative (GRI), the Carbon Disclosure Project (CDP), or the Sustainable Apparel Coalition to standardize ESG reporting, share information, and drive sector-wide improvements in environmental and social performance.

Supply Chain Collaboration

Working closely with your suppliers and other supply chain partners can help you address ESG risks and opportunities throughout your value chain. By fostering dialogue and collaboration, you can identify areas for improvement, share best practices, and jointly develop solutions to enhance the sustainability performance of your products and services. For instance, you can collaborate with suppliers to implement responsible sourcing practices, reduce packaging waste, or improve labor conditions in your supply chain.

Community Engagement and Local Partnerships

Engaging with local communities and partnering with grassroots organizations can help you address the social and environmental dimensions of ESG sustainability in a more contextualized and inclusive manner. By collaborating with community-based organizations, you can identify local needs, develop tailored solutions, and create shared value. For example, you can partner with local NGOs to support education, healthcare, or livelihood initiatives, promote local biodiversity conservation, or implement community-based natural resource management projects.

Embracing cross-sector collaboration and partnerships will allow. You to harness the collective expertise, resources, and influence of diverse stakeholders to address the complex ESG challenges that transcend traditional sector boundaries. In doing so, you will not only enhance your own sustainability performance but also contribute to the broader global effort to create a more sustainable and resilient future.

Lessons Learned and Success Stories

Learning from success stories and understanding the lessons learned from various ESG initiatives can provide valuable insights to help you improve your sustainability performance and drive positive change. Here, we present some practical examples that illustrate the importance of embracing ESG principles and the benefits that can be derived from implementing effective sustainability strategies.

Lesson 1: Embed ESG into your corporate culture

A key lesson from successful organizations is the importance of embedding ESG principles into the corporate culture. By prioritizing sustainability at the highest level of decision-making and involving employees at all levels, you can create a strong foundation for driving continuous improvement and innovation in ESG performance. For example, Unilever's Sustainable Living Plan showcases the company's commitment to embedding sustainability across its value chain, which has led to significant reductions in environmental impacts, as well as the development of products that meet the evolving needs of conscious consumers.

Lesson 2: Set ambitious targets and measure progress

Setting ambitious ESG targets and regularly measuring progress can help you maintain focus, drive improvements, and demonstrate your commitment to stakeholders. For instance, Ørsted, a Danish renewable energy company, set an ambitious goal to become carbon-neutral by 2025 and completely transform its business from fossil fuels to renewables. By committing to this vision and tracking progress, Ørsted has become a global leader in renewable energy, with a significant positive impact on both the environment and its financial performance.

Lesson 3: Engage with stakeholders and be transparent

Engaging with stakeholders and being transparent about your ESG performance is crucial for building trust and ensuring that your initiatives are aligned with their expectations. For example, the clothing company Patagonia actively engages with customers, suppliers, and NGOs to address issues related to ethical supply chains, fair labor practices, and environmental conservation. Their transparency in reporting progress and sharing challenges has earned them a reputation as a responsible and sustainable brand.

Lesson 4: Foster innovation and invest in sustainable technologies

Investing in sustainable technologies and fostering a culture of innovation can help you stay ahead of the curve and create long-term value. Tesla, for instance, has been a pioneer in electric vehicle (EV) technology, not only revolutionizing the automotive industry but also contributing to the global transition toward clean energy. By investing in R&D and embracing innovative solutions, Tesla has become a market leader and demonstrated the business potential of sustainable technologies.

Lesson 5: Adapt to regulatory changes and seize opportunities

Successful organizations are proactive in adapting to regulatory changes and seizing the opportunities presented by evolving ESG frameworks. An example is Bank of America, which has integrated ESG into its core business strategy and risk management processes. This has allowed the bank to capitalize on the growing demand for sustainable finance and position itself as a leader in ESG investing.

Studying these lessons learned and success stories will allow you to gain valuable insights and inspiration for enhancing your own ESG performance. Understanding the strategies and approaches that have enabled these organizations to excel in sustainability can help you identify opportunities for improvement, drive innovation, and create long-term value for your business and stakeholders.

Module 8 Exam Questions

1. Which of the following is an example of an environmental criterion in ESG sustainability?

a) Biodiversity conservation

b) Employee health and safety

c) Executive compensation

d) Shareholder rights

2. Which aspect of social sustainability focuses on employee well-being and a safe working environment?

a) Human rights and labor standards

b) Employee health, safety, and well-being

c) Diversity, equity, and inclusion

d) Community engagement and development

3. What is the main purpose of an ESG reporting framework?

a) To measure a company's financial performance

b) To assess a company's environmental impact

c) To standardize and communicate a company's ESG performance

d) To evaluate a company's product quality

4. Which sector can contribute to ESG sustainability by focusing on responsible forest management and resource efficiency?

a) Technology sector

b) Forestry and pulp and paper sector

c) Professional services sector

d) Maritime and shipping sector

5. What is a key benefit of cross-sector collaboration in ESG sustainability?

a) Reduced competition

b) Enhanced marketing efforts

c) Accelerated progress towards sustainability goals

d) Simplified regulatory compliance

6. Which renewable energy company transformed its business from fossil fuels to renewables and set an ambitious goal to become carbon-neutral by 2025?

a) ExxonMobil

b) Shell

c) Ørsted

d) Chevron

7. Which clothing company is known for engaging with stakeholders and being transparent about its ESG performance, especially in ethical supply chains and fair labor practices?

a) Nike

b) H&M

c) Patagonia

d) Zara

8. Which of the following is an example of a governance criterion in ESG sustainability?

a) Climate change and greenhouse gas emissions

b) Resource management and waste reduction

c) Executive compensation and performance management

d) Employee health, safety, and well-being

9. What is the primary focus of the waste management sector in terms of ESG sustainability?

a) Employee health and safety

b) Corporate governance structures and policies

c) Waste reduction and circular economy principles

d) Board composition and diversity

10. What is an example of a social criterion in ESG sustainability?

a) Air and water pollution prevention

b) Talent attraction, retention, and development

c) Sustainable procurement and supply chain management

d) Environmental management systems

11. Which initiative can help standardize ESG reporting and share information to drive sector-wide improvements in environmental and social performance?

a) Global Reporting Initiative (GRI)

b) International Monetary Fund (IMF)

c) World Trade Organization (WTO)

d) World Health Organization (WHO)

12. What is a key lesson learned from successful ESG initiatives?

a) Focus on short-term gains

b) Keep ESG strategies separate from core business activities

c) Embed ESG principles into corporate culture

d) Avoid setting ambitious ESG targets

13. Which of the following is an example of a public-private partnership (PPP)?

a) Collaboration between a government entity and a private company to develop sustainable infrastructure

b) A private company investing in its own renewable energy project

c) Two private companies partnering to create a joint venture

d) A government agency collaborating with another government agency

14. In which sector can ESG initiatives focus on reducing the environmental impact of shipping operations and promoting responsible supply chain management?

a) Retail sector

b) Construction sector

c) Maritime and shipping sector

d) Pharmaceutical sector

15. What is the primary goal of environmental sustainability in the context of ESG?

a) To minimize negative environmental impacts and protect natural resources

b) To maximize shareholder value

c) To ensure employee well-being

d) To promote good corporate governance

16. Which of the following is a key component of governance sustainability?

a) Biodiversity and ecosystem preservation

b) Community engagement and development

c) Business ethics and anti-corruption

d) Sustainable procurement and supply chain management

17. What is the main objective of ESG integration in investment and finance?

a) To prioritize short-term profits

b) To incorporate ESG factors into investment decision-making

c) To focus solely on environmental aspects of investments

d) To ignore social and governance factors

18. Which financial instrument is specifically designed to fund environmentally-friendly projects?

a) Corporate bonds

b) Green bonds

c) Municipal bonds

d) Treasury bonds

19. What is the primary purpose of engaging in collaborative research and development with academia and research institutions for ESG sustainability?

a) To reduce marketing expenses

b) To foster innovation and develop new sustainable solutions

c) To improve corporate governance

d) To solely focus on financial performance

20. Which of the following is an essential aspect of ESG reporting and disclosure?

a) Materiality assessment and stakeholder engagement

b) Focusing only on positive aspects of ESG performance

c) Ignoring data quality and accuracy

d) Communicating ESG performance only to internal stakeholders

Answers:

1. a
2. b
3. c
4. b
5. c
6. c
7. c
8. c
9. c
10. b
11. a
12. c
13. a
14. c
15. a
16. c
17. b
18. b
19. b
20. a

Module 9. The Future of ESG Sustainability

Introduction to The Future of ESG Sustainability

Welcome to Module 9 on The Future of ESG Sustainability. In this module, you will explore the emerging trends, challenges, and opportunities that lie ahead in the ever-evolving world of Environmental, Social, and Governance (ESG) sustainability. As businesses and investors increasingly recognize the importance of ESG considerations, staying ahead of the curve is essential for long-term success and resilience.

You will gain insights into how ESG sustainability is expected to evolve in the coming years, as well as the role that technology and innovation will play in driving sustainable development. By understanding the future landscape of ESG, you can better anticipate changes and proactively adapt your strategies to meet the needs of an increasingly interconnected and environmentally-conscious global market.

Through practical examples and real-world applications, this module will help you envision the future of ESG sustainability and the role your organization can play in shaping a more sustainable world. We will discuss how the policy landscape and regulatory developments may impact ESG performance and reporting, and how businesses can leverage these changes to create value for all stakeholders.

As we delve into the future of ESG sustainability, you will be equipped with the knowledge and foresight to navigate the emerging trends and challenges, ensuring that your organization remains at the forefront of sustainable business practices and is well-prepared to capitalize on the opportunities that lie ahead.

Emerging ESG Trends and Challenges

As you navigate the future of ESG sustainability, it is crucial to be aware of the emerging trends and challenges that will shape the landscape. In this section, we will discuss several key developments and their implications for your ESG strategy.

- **Climate Crisis Intensification:** As the impacts of climate change become more severe, the urgency to address greenhouse gas emissions and adapt to the changing environment increases. Your business will need to develop robust climate strategies, such as setting science-based targets and investing in low-carbon technologies, to contribute to the global effort in mitigating climate change.

- **Circular Economy:** The transition from a linear to a circular economy, where resources are kept in use for as long as possible, is gaining momentum. This shift requires you to reevaluate your production processes, supply chain management, and product design to minimize waste, enhance resource efficiency, and capitalize on new business opportunities.

- **Social Issues and Inequalities:** Growing awareness of social issues, such as income inequality, racial injustice, and gender disparity, is prompting businesses to place a stronger focus on social sustainability. You will need to examine your workforce policies, community engagement, and diversity and inclusion initiatives to ensure that you are fostering an equitable and inclusive environment.

- **Digital Transformation and Data Privacy:** The increasing reliance on digital technologies brings opportunities for efficiency and innovation but also raises concerns about data privacy and security. It is essential for you to invest in robust cybersecurity measures and maintain transparency in data handling practices to protect your stakeholders' interests and build trust.

- **Supply Chain Resilience:** Global supply chain disruptions, such as those caused by the COVID-19 pandemic, have highlighted the need for businesses to build more resilient and sustainable supply chains. To achieve this, you should consider diversifying suppliers, implementing traceability systems, and collaborating with your suppliers on ESG performance improvements.
- **Stakeholder Activism and Expectations:** Stakeholders, including investors, employees, and consumers, are increasingly demanding higher ESG standards from businesses. To meet these expectations, you must prioritize transparent ESG reporting, engage with stakeholders, and continuously refine your ESG strategy based on their feedback.
- **Regulatory Landscape Evolution:** Governments worldwide are strengthening ESG-related regulations, such as carbon pricing, mandatory ESG reporting, and human rights due diligence. You should closely monitor these regulatory developments and proactively adapt your ESG approach to ensure compliance and mitigate potential risks.

Staying informed about these emerging trends and challenges will allow you to better prepare your business for the future of ESG sustainability, proactively address risks, and seize opportunities for growth and value creation.

The Role of Technology and Innovation

Technology and innovation play a critical role in advancing ESG sustainability. They enable you to develop new solutions, improve efficiency, and better manage risks associated with environmental, social, and governance factors. In this section, we will explore some key technologies and innovations that can enhance your ESG performance.

- **Renewable Energy Technologies:** The transition to clean energy is essential for addressing climate change. As you seek to reduce your carbon footprint, investing in renewable energy technologies, such as solar panels, wind turbines, and energy storage solutions, can help you minimize greenhouse gas emissions and lower your reliance on fossil fuels.

- **Artificial Intelligence (AI) and Machine Learning:** AI and machine learning can be powerful tools for analyzing large volumes of ESG data, enabling you to identify trends, risks, and opportunities more effectively. You can leverage these technologies to optimize resource allocation, monitor supply chain performance, and improve decision-making processes.

- **Internet of Things (IoT):** IoT devices, such as sensors and smart meters, can help you monitor and manage various aspects of your operations in real-time. By integrating IoT into your facilities, you can track energy consumption, water usage, and waste generation more accurately, enabling you to identify inefficiencies and optimize resource management.

- **Blockchain:** Blockchain technology offers increased transparency, traceability, and security in managing ESG data. You can use blockchain to track and verify the origin of raw materials, ensure ethical labor practices in your supply chain, and enhance the credibility of your ESG reporting.

- **Green Technologies:** Innovative green technologies, such as carbon capture and storage, biodegradable packaging, and water purification systems, can help you address specific environmental challenges. By adopting these solutions, you can reduce your environmental footprint, mitigate risks, and demonstrate your commitment to sustainability.

- **Remote Work and Digital Collaboration:** The shift to remote work and digital collaboration tools, accelerated by the COVID-19 pandemic, has the potential to reduce your carbon emissions and enhance employee well-being. By embracing these tools and promoting a flexible work culture, you can create a more sustainable and inclusive workplace.

- **ESG Fintech Solutions:** Financial technology (fintech) solutions are increasingly being developed to address ESG challenges and opportunities. You can leverage these tools to improve ESG data management, reporting, and analysis, as well as streamline sustainable investment processes.

As you navigate the future of ESG sustainability, embracing technology and innovation can help you stay ahead of the curve, drive continuous improvement, and create lasting value for your stakeholders. By harnessing the power of these emerging solutions, you can effectively address ESG challenges, capitalize on opportunities, and foster a more sustainable future.

Policy Landscape and Regulatory Developments

The policy landscape and regulatory developments related to ESG are rapidly evolving, with governments and international organizations introducing new measures to drive sustainable practices and ensure greater corporate accountability. In this section, we will discuss some key developments that may affect your ESG strategy and performance.

- **Mandatory ESG Reporting:** Many countries are moving towards mandatory ESG reporting for companies, requiring you to disclose information on environmental, social, and governance factors in a standardized and transparent manner. Being proactive in your ESG reporting and aligning your disclosure with recognized frameworks, such as the Global Reporting Initiative (GRI) or the Sustainability Accounting Standards Board (SASB), can help you meet these requirements and demonstrate your commitment to sustainability.

- **Carbon Pricing and Emissions Regulations:** Governments are increasingly implementing carbon pricing mechanisms, such as carbon taxes or cap-and-trade systems, to incentivize businesses to reduce their greenhouse gas emissions. You should closely monitor these developments, assess the financial implications for your operations, and explore opportunities to minimize your emissions through energy efficiency improvements and investments in renewable energy.

- **Human Rights and Labor Regulations:** Regulations related to human rights and labor standards are becoming more stringent, aiming to prevent human rights abuses and promote fair labor practices in global supply chains. You should ensure your policies and practices align with these standards, such as the United Nations Guiding Principles on Business and Human Rights, and engage with your suppliers to encourage compliance throughout your supply chain.

- **Environmental Regulations:** Governments are introducing stricter environmental regulations to protect natural resources and ecosystems, such as restrictions on single-use plastics, water management guidelines, and biodiversity conservation measures. It is crucial for you to stay informed about these regulatory changes, evaluate their impact on your business, and develop strategies to minimize your environmental footprint and maintain compliance.

- **Sustainable Finance Regulations:** Financial regulators are increasingly integrating ESG considerations into their frameworks, leading to new requirements for financial institutions and investors. These regulations may affect your access to capital and financing opportunities, making it essential for you to understand and address the ESG risks and opportunities in your business and engage with investors to demonstrate your commitment to sustainability.

- **Climate Risk Disclosure:** Regulatory bodies are recognizing the financial risks associated with climate change and are developing guidelines for climate risk disclosure. By aligning your reporting with frameworks such as the Task Force on Climate-related Financial Disclosures (TCFD), you can demonstrate your preparedness for climate-related risks and improve your access to capital from sustainability-focused investors.

Keeping abreast of the changing policy landscape and regulatory developments is essential for managing ESG risks and opportunities in your business. By proactively adapting your ESG strategy and practices, you can ensure compliance, enhance your reputation, and drive long-term value creation for all stakeholders.

Shaping a Sustainable Future through ESG

As you navigate the future of ESG sustainability, your actions and decisions will play a crucial role in shaping a more sustainable future for your business, society, and the planet. In this section, we will discuss some strategies and practical examples that can help you contribute to a sustainable future through effective ESG management.

- **Align with Global Sustainability Goals:** By aligning your ESG strategy with internationally recognized goals, such as the United Nations Sustainable Development Goals (SDGs), you can ensure that your efforts contribute to broader sustainability objectives. Identify the SDGs most relevant to your business and develop targeted initiatives to address those goals, creating a positive impact on society and the environment.

- **Drive Innovation:** Embrace technology and innovation to develop new solutions that address pressing ESG challenges. For example, you might invest in research and development to create eco-friendly products or collaborate with startups to explore new approaches to resource management and waste reduction. By driving innovation, you can generate value for your business and contribute to a sustainable future.

- **Foster a Culture of Sustainability:** Cultivating a culture of sustainability within your organization is essential for driving long-term ESG performance. Encourage your employees to participate in sustainability initiatives, provide training on ESG topics, and integrate sustainability considerations into your corporate values and decision-making processes. By embedding sustainability into your organizational culture, you can inspire your workforce to contribute to a sustainable future.

- **Collaborate with Stakeholders:** Engaging with stakeholders, such as suppliers, customers, investors, and local communities, is crucial for identifying shared ESG priorities and fostering collaboration. Establish partnerships and joint initiatives that address common sustainability challenges, such as reducing supply chain emissions, promoting circular economy practices, or supporting community development projects.

- **Set Ambitious ESG Targets:** Establish clear, ambitious, and time-bound ESG targets to drive continuous improvement and measure your progress towards a sustainable future. Regularly assess your performance against these targets and adjust your strategy as needed to ensure you remain on track to achieve your sustainability objectives.

- **Advocate for Sustainable Policies:** Use your influence to advocate for progressive ESG policies and regulations that promote sustainable development. By engaging with policymakers and participating in industry associations, you can help shape a policy landscape that supports responsible business practices and creates a level playing field for all market participants.

Implementing these strategies and taking a proactive approach to ESG management will allow you to contribute to shaping a sustainable future for your business, society, and the planet. As you work towards a more sustainable future, you will not only mitigate risks and capitalize on opportunities but also create lasting value for all stakeholders, ensuring your business remains resilient and competitive in the long term.

The EU's Corporate Sustainability Reporting Directive (CSRD)

Corporate sustainability reporting is an essential aspect of ESG management, enabling you to disclose information on social and environmental risks, opportunities, and impacts related to your business activities. In the European Union (EU), specific rules and directives govern sustainability reporting for large companies and listed companies, ensuring transparency and helping stakeholders assess your sustainability performance.

- **EU Directives on Sustainability Reporting:** The Non-Financial Reporting Directive (NFRD) established the initial framework for large public-interest companies with over 500 employees to disclose information on environmental, social, and governance matters. However, the recently introduced Corporate Sustainability Reporting Directive (CSRD) expands the scope, including a broader range of companies (approximately 50,000) and requiring more detailed reporting.

- **Reporting Requirements:** Under the CSRD, you will need to report on various ESG aspects, including environmental matters, social issues and employee treatment, human rights, anti-corruption and bribery, and diversity on company boards. These requirements help ensure that investors and other stakeholders have access to relevant information to assess investment risks and opportunities related to sustainability.

- **European Sustainability Reporting Standards (ESRS):** Companies subject to the CSRD will have to report according to the ESRS. These standards, developed by the European Financial Reporting Advisory Group (EFRAG), are tailored to EU policies while also contributing to international standardization initiatives. The first set of ESRS is expected to be adopted by mid-2023.

- **Auditing and Digitalization:** The CSRD mandates that companies have their sustainability information audited and includes provisions for digitalizing sustainability information, ensuring greater transparency and accuracy in reporting.
- **Timeline:** The CSRD entered into force on January 5, 2023, and the first companies will apply the new rules for the 2024 financial year, with reports published in 2025.

Practical Guidance: As you prepare for corporate sustainability reporting under the CSRD, consider the following practical steps:

1. Identify the specific ESG aspects you need to report on, based on the new directive and relevant ESRS.
2. Establish processes for collecting, analyzing, and managing ESG data, ensuring its accuracy and reliability.
3. Engage with internal and external stakeholders to gather information on your ESG risks, opportunities, and performance.
4. Collaborate with an external auditor to review and verify your sustainability information.
5. Develop a comprehensive and transparent sustainability report that aligns with the CSRD requirements and ESRS.

Corporate sustainability reporting is a critical component of your ESG management and helps you demonstrate your commitment to sustainable development. By adhering to the EU's directives and standards, you can foster transparency, build trust with stakeholders, and contribute to a more sustainable future.

Module 9 Exam Questions

1. Which of the following is an aspect of environmental sustainability?

a) Human Rights and Labor Standards

b) Corporate Governance Structures and Policies

c) Climate Change and Greenhouse Gas Emissions

d) Community Engagement and Development

2. Which aspect of social sustainability focuses on a company's responsibility towards its customers?

a) Product Responsibility and Customer Safety

b) Talent Attraction, Retention, and Development

c) Executive Compensation and Performance Management

d) Environmental Management Systems

3. Which ESG component includes board composition and diversity?

a) Environmental Sustainability

b) Social Sustainability

c) Governance Sustainability

d) ESG Reporting and Disclosure

4. Which of the following is a key element of ESG reporting?

a) Materiality Assessment and Stakeholder Engagement

b) Developing an ESG Strategy and Roadmap

c) ESG Risk and Opportunity Assessment

d) Biodiversity and Ecosystem Preservation

5. Green bonds are an example of:

a) ESG Criteria and Key Performance Indicators

b) Responsible Investing and ESG Factors

c) Corporate Governance Structures and Policies

d) Sustainable Procurement and Supply Chain Management

6. The United Nations Sustainable Development Goals (SDGs) are related to:

a) ESG Sustainability

b) Corporate Governance Structures and Policies

c) Executive Compensation and Performance Management

d) Talent Attraction, Retention, and Development

7. Which of the following is an example of environmental sustainability?

a) Pollution Prevention and Control

b) Human Rights and Labor Standards

c) Executive Compensation and Performance Management

d) Shareholder Rights and Engagement

8. A key aspect of governance sustainability is:

a) Regulatory Compliance and Risk Management

b) Talent Attraction, Retention, and Development

c) Resource Management and Waste Reduction

d) Diversity, Equity, and Inclusion

9. A company's focus on employee health, safety, and well-being is part of:

a) Environmental Sustainability

b) Social Sustainability

c) Governance Sustainability

d) ESG Reporting and Disclosure

10. ESG integration in portfolio management is most relevant to:

a) ESG in Investment and Finance

b) Environmental Sustainability

c) Social Sustainability

d) Governance Sustainability

11. The Corporate Sustainability Reporting Directive (CSRD) applies to companies in:

a) The United States

b) The European Union

c) Asia

d) Africa

12. What is the purpose of the European Sustainability Reporting Standards (ESRS)?

a) To guide companies in ESG risk assessment

b) To regulate ESG integration in portfolio management

c) To standardize corporate sustainability reporting in the EU

d) To develop new ESG investment products

13. In which year did the Corporate Sustainability Reporting Directive (CSRD) enter into force?

a) 2021

b) 2022

c) 2023

d) 2024

14. Sustainable Procurement and Supply Chain Management is a component of:

a) Environmental Sustainability

b) Social Sustainability

c) Governance Sustainability

d) ESG Reporting and Disclosure

15. ESG Reporting Frameworks and Standards are associated with:

a) ESG Reporting and Disclosure

b) Environmental Sustainability

c) Social Sustainability

d) Governance Sustainability

16. Which of the following is part of social sustainability?

a) Climate Change and Greenhouse Gas Emissions

b) Corporate Governance Structures and Policies

c) Biodiversity and Ecosystem Preservation

d) Diversity, Equily, and Inclusion

17. What is the primary purpose of ESG reporting?

a) To evaluate the sustainability performance of companies

b) To assess a company's financial performance

c) To compare different companies' management structures

d) To ensure that companies comply with environmental regulations

18. Which of the following is an example of a governance sustainability issue?

a) Water consumption and efficiency

b) Employee training and development programs

c) Board composition and diversity

d) Waste reduction and recycling

19. In the context of ESG, materiality assessment is used to:

a) Determine the financial value of a company

b) Identify the most relevant ESG issues for a company

c) Compare the performance of different companies

d) Measure the environmental impact of a company's operations

20. Continuous improvement and innovation in ESG are related to:

a) ESG Integration and Strategy

b) Environmental Sustainability

c) Social Sustainability

d) Governance Sustainability

Answers:

1. c
2. a
3. c
4. a
5. b
6. a
7. a
8. a
9. b
10. a
11. b
12. c
13. c
14. a
15. a
16. d
17. a
18. c
19. b
20. a

Module 10. Building a Career in Sustainability

Introduction to Career Development and Certification

The field of sustainability is rapidly growing, and there is a high demand for professionals with the skills and knowledge to tackle the challenges of creating a more sustainable future. If you are interested in building a career in sustainability, it is important to understand the various aspects of the field, as well as the skills and qualifications needed to succeed.

In this module 10 on Building a career in Sustainability, we will provide you with practical information and advice on how to pursue a successful career in sustainability. We will cover topics such as identifying transferable skills, networking and industry engagement, building a strong sustainability portfolio, pursuing job opportunities, and continuing your lifelong learning and career growth.

We recommend taking the certification program offered by the International Institute of Executive Careers (IIEC), available at TheBestMBA.org, to further enhance your knowledge and credibility in the field of sustainability. This program is designed to provide you with a comprehensive understanding of sustainability as well as practical tools and techniques for implementing sustainability initiatives.

By the end of this module, you will have a better understanding of the steps needed to build a successful career in sustainability, as well as the resources and tools available to help you achieve your goals. We strongly encourage you to consider the IIEC certification program as a valuable investment in your future career.

Leveraging Existing Skills and Knowledge

As you embark on your journey to build a successful career in sustainability, one of the first steps is to recognize and leverage your existing skills and knowledge. Whether you come from a business, engineering, or social science background, there's a good chance that your expertise can be applied to the rapidly evolving field of sustainability. By identifying your transferable skills and knowledge, you can better position yourself as a valuable asset in the sustainability sector.

Transferable skills are those that can be used across various industries and roles. They are often referred to as "soft skills" and include abilities such as communication, problem-solving, leadership, and project management. In the context of sustainability, these skills can help you effectively collaborate with cross-functional teams, develop strategies, and execute sustainability initiatives.

For instance, if you have a strong background in data analysis, you can apply this skill to analyze environmental, social, and governance (ESG) data, helping organizations make informed decisions and set sustainability goals. If you have experience in marketing or communications, you can use your storytelling skills to engage stakeholders and raise awareness of sustainability initiatives, both internally and externally.

Similarly, your knowledge in specific domains can also be valuable in sustainability roles. For example, if you have a background in supply chain management, you can help companies identify opportunities for reducing waste and improving the sustainability of their operations. If you have experience in finance, you can contribute to the development of green financing strategies and assess the ESG performance of investment portfolios.

To leverage your existing skills and knowledge, it's crucial to understand how they relate to sustainability challenges and opportunities. Start by researching the most pressing issues in your industry and explore how your expertise can contribute to addressing them. Look for case studies, articles, or reports that highlight the use of similar skills and knowledge in sustainability roles, and reflect on how you can apply your unique strengths to make a positive impact in this field. By doing so, you'll be well on your way to building a career in sustainability as a game changer.

Networking and Industry Engagement

As you forge your path in the sustainability sector, networking and industry engagement are crucial components to your success. These activities allow you to connect with like-minded professionals, learn from their experiences, and stay up-to-date with the latest trends and developments in the field. Building a strong professional network can open doors to exciting opportunities, facilitate collaborations, and provide invaluable insights to help you advance your career in sustainability.

Sustainability conferences, workshops, and webinars are excellent platforms for networking and industry engagement. These events bring together professionals from various sectors and offer a wealth of knowledge through presentations, panel discussions, and workshops. By attending such events, you not only gain exposure to the latest research and best practices in sustainability but also have the opportunity to interact with industry leaders, academics, and fellow enthusiasts. Some notable examples of sustainability conferences include the GreenBiz Forum, the Sustainable Brands Conference, and the Global Climate Action Summit.

In addition to attending events, consider joining sustainability-focused organizations and online communities. These groups provide ongoing support, resources, and networking opportunities that can help you stay connected with your peers and learn from their experiences. Some examples of such organizations are the Association for Sustainability Professionals, the International Society of Sustainability Professionals, and the Global Reporting Initiative (GRI) Community. Online communities, such as LinkedIn groups or sustainability-focused forums, can also offer valuable connections, resources, and discussions.

Actively engaging in networking and industry events will allow you to expand your professional circle and tap into a wealth of collective wisdom. As you participate in these activities, remember to approach them with a genuine curiosity and willingness to learn. Be open to new ideas, share your own insights, and foster meaningful relationships that can support your career growth in the dynamic world of sustainability.

Building a Strong Sustainability Portfolio

As you pursue a career in sustainability, it's essential to showcase your expertise and demonstrate your commitment to the field through a strong sustainability portfolio. A well-rounded portfolio highlights your achievements, experiences, and learnings, making it an indispensable tool for capturing the attention of potential employers and collaborators.

One way to build your sustainability portfolio is by working on personal projects related to your area of interest. For instance, you could create a blog or podcast focused on sustainable business practices, conduct research on renewable energy solutions, or develop a tool to help individuals reduce their carbon footprint. Personal projects allow you to deepen your understanding of sustainability issues and demonstrate your ability to apply your skills and knowledge in practical settings.

Volunteering for sustainability initiatives is another excellent way to expand your portfolio. Look for opportunities to support non-profit organizations, community projects, or events that align with your sustainability goals. For example, you could volunteer with organizations like The Nature Conservancy, World Wildlife Fund, or local environmental groups to gain hands-on experience in habitat restoration, conservation, or community outreach. Volunteering not only enhances your skillset but also showcases your dedication to making a positive impact.

Contributing to open-source projects is yet another opportunity to demonstrate your expertise in sustainability. Many organizations, researchers, and developers use open-source platforms to collaborate on sustainability-related projects. By contributing to these projects, you can help advance solutions to pressing sustainability challenges while showcasing your technical skills and collaborative abilities.

Finally, it's crucial to document your achievements, experiences, and learnings in a professional portfolio or LinkedIn profile. These platforms serve as a digital showcase of your sustainability expertise and can help you connect with potential employers or collaborators. Be sure to include detailed descriptions of your projects, the skills you applied, and the impact you made. Regularly updating your portfolio with new experiences and accomplishments demonstrates your ongoing commitment to sustainability and personal growth.

Building a strong sustainability portfolio will allow you to be well-equipped to showcase your expertise, attract opportunities, and drive positive change in the sustainability sector.

Pursuing Job Opportunities in Sustainability

As you seek to build a career in sustainability, it's essential to research and pursue job opportunities that align with your skills, interests, and values. The growing importance of sustainability has led to an increasing number of roles within companies, non-profit organizations, and governmental agencies, offering diverse opportunities for professionals with various backgrounds and expertise.

To find sustainability-focused roles, start by researching companies and organizations that prioritize sustainability in their operations or mission. Look for job postings on their websites, as well as on dedicated sustainability job boards such as GreenBiz, SustainableBusiness.com, or the Environmental Career Opportunities (ECO) job board. Networking events, industry conferences, and online communities can also provide valuable leads on job openings.

Sustainability roles can be found across a wide range of industries and sectors. Some examples of positions include Sustainability Manager, Corporate Social Responsibility (CSR) Specialist, Environmental Analyst, and Renewable Energy Consultant. These roles often involve working on ESG strategy, data analysis, stakeholder engagement, and policy development, among other responsibilities. When applying for these positions, make sure to highlight your relevant skills, experiences, and achievements in your application materials and during interviews.

Another option is to consider starting as a sustainability consultant. As a consultant, you can offer your expertise to businesses and organizations looking to incorporate ESG practices into their operations. This can involve assessing their sustainability performance, developing strategies for improvement, and implementing sustainable solutions. To establish yourself as a sustainability consultant, it's crucial to build a strong portfolio showcasing your expertise, as well as to develop a network of clients and collaborators. You can begin by offering your services on a project basis or collaborating with established consulting firms specializing in sustainability.

Pursuing job opportunities in sustainability requires persistence, adaptability, and a genuine passion for driving positive change. By staying informed about the latest industry trends, continuously expanding your skillset, and leveraging your network, you'll be well-prepared to seize exciting opportunities and make a meaningful impact in the sustainability sector.

Learning and Career Growth

In the ever-evolving field of sustainability, lifelong learning and career growth are essential components of success. As new challenges emerge and innovative solutions are developed, staying informed and continuously expanding your skills will not only help you remain relevant but also contribute to your personal and professional growth in the sector.

One way to stay informed about the latest developments, innovations, and best practices in sustainability is to regularly read industry reports, articles (I recommend you this website: https://www.esgtoday.com), and case studies from reputable sources. Follow organizations, thought leaders, and publications that focus on sustainability issues to ensure you're up-to-date with the most recent information. Additionally, subscribing to newsletters or podcasts can provide you with a steady stream of insights and perspectives from experts in the field.

Continuously seeking opportunities for professional development is also crucial for enhancing your skills and marketability. Further education, certifications, or training programs can provide you with a deeper understanding of sustainability concepts and specialized skills in various areas. For instance, you might consider pursuing a master's degree in sustainability management, earning a certification in environmental management systems (such as ISO 14001), or attending training courses on topics like ESG reporting or green finance.

Additionally, participating in workshops, webinars, and conferences can offer opportunities to learn about new developments and engage with experts in your field. Many of these events also provide continuing education credits, which can be valuable for maintaining professional certifications or memberships in sustainability organizations.

Finally, don't underestimate the power of learning from your peers and mentors. Engaging in conversations, sharing experiences, and seeking advice from others in the sustainability sector can offer unique insights and perspectives to help you grow professionally. Networking events, online communities, and workplace mentorship programs can provide invaluable opportunities to connect with and learn from other sustainability professionals.

Embracing lifelong learning and actively seeking opportunities for career growth will allow you to be well-equipped to navigate the dynamic landscape of sustainability and make a lasting impact in your chosen field.

Module 10 Exam Questions

1. What is a key component of environmental sustainability?

a) Employee health, safety, and well-being

b) Corporate governance structures and policies

c) Climate change and greenhouse gas emissions

d) Responsible investing and ESG factors

2. What is the main focus of social sustainability?

a) Human rights and labor standards

b) Pollution prevention and control

c) Regulatory compliance and risk management

d) Resource management and waste reduction

3. Which of these aspects falls under governance sustainability?

a) Biodiversity and ecosystem preservation

b) Community engagement and development

c) Sustainable procurement and supply chain management

d) Executive compensation and performance management

4. What is the main purpose of ESG reporting and disclosure?

a) To attract top talent and improve employee retention

b) To communicate an organization's ESG performance to stakeholders

c) To develop a company's marketing strategy

d) To expand a company's product offerings

5. What is the primary goal of ESG integration and strategy?

a) To increase an organization's profits

b) To improve employee morale

c) To gain a competitive advantage in the market

d) To identify and assess ESG risks and opportunities

6. Green bonds are an example of which category?

a) Environmental sustainability

b) Social sustainability

c) Governance sustainability

d) ESG in investment and finance

7. In which module can you find information about industry-specific ESG initiatives and innovations?

a) Module 6

b) Module 7

c) Module 8

d) Module 9

8. What is the main focus of the module "10. Building a Career in Sustainability for Game Changers"?

a) Discovering your passion for sustainability

b) Learning about the history of sustainability

c) Exploring different industries and their sustainability efforts

d) Encouraging individuals to take action and build a career in sustainability

9. Which skill is considered a transferable skill that can be applied to a career in sustainability?

a) Communication

b) Geology

c) Thermodynamics

d) Tax law

10. Which of the following is an example of a sustainability-focused organization?

a) The American Marketing Association

b) The Association for Sustainability Professionals

c) The National Basketball Association

d) The American Bar Association

11. What is a key benefit of attending sustainability conferences, workshops, and webinars?

a) Building your professional network

b) Improving your cooking skills

c) Learning a new language

d) Enhancing your fashion sense

12. Which of the following can help build a strong sustainability portfolio?

a) Watching TV shows about sustainability

b) Reading fiction books

c) Volunteering for sustainability initiatives

d) Playing video games

13. What is the primary goal of working as a sustainability consultant?

a) To help businesses and organizations incorporate ESG practices into their operations

b) To increase personal income

c) To travel the world

d) To develop new products for the market

14. Which of the following is an example of a sustainability-focused job board?

a) GreenBiz

b) LinkedIn

c) Indeed

d) Monster

15. What is the main purpose of pursuing further education, certifications, or training programs in sustainability?

a) To impress your friends and family

b) To enhance your skills and marketability

c) To become a famous influencer

d) To win awards and recognition

16. Which of the following is an example of a sustainability certification?

a) ISO 14001

b) Six Sigma

c) Project Management Professional (PMP)

d) Certified Public Accountant (CPA)

17. Which of the following is an example of a personal project that can help build your sustainability portfolio?

a) Creating a blog focused on sustainable business practices

b) Hosting a cooking show

c) Writing a novel

d) Designing a fashion collection

18. In which of the following sectors can sustainability roles be found?

a) Education

b) Healthcare

c) Manufacturing

d) All of the above

19. What is a potential benefit of attending workshops, webinars, and conferences focused on sustainability?

a) Practicing your public speaking skills

b) Learning about new developments in the field

c) Enhancing your musical abilities

d) Exploring new vacation destinations

20. Why is networking an important aspect of building a career in sustainability?

a) It helps you find job opportunities

b) It allows you to meet new friends

c) It improves your artistic skills

d) It boosts your self-esteem

Answers:

1. c
2. a
3. d
4. b
5. d
6. d
7. c
8. d
9. a
10. d
11. a
12. c
13. a
14. a
15. b
16. a
17. a
18. d
19. b
20. a

Conclusions and Next Steps

As we reach the end of this book, it is clear that the need for sustainability is greater than ever before, and its impact on our personal and professional lives is undeniable. Throughout this journey, we have explored the challenges we face as a global community, the opportunities for creating a sustainable future, and the role that individuals and organizations play in achieving this goal.

We have delved into the strategies for cultivating a resilient mindset, navigating the ethical landscape, and developing sustainability leadership and organizational readiness. The future of work will be shaped by the seamless integration of sustainability principles with human creativity, social responsibility, and adaptability.

By understanding the strengths and limitations of sustainable practices, we can harness their power to create a sustainable and equitable future for all. However, to truly make a difference, it is essential for each individual and organization to take proactive steps towards embracing sustainable practices.

As a game changer, you have the power to make a lasting impact on your business and the world. Here are the next steps you can take to embrace sustainability and create positive change:

1. **Enroll in a sustainability program:** Gain in-depth knowledge, practical tools, and valuable insights from a comprehensive sustainability curriculum. This will help you thrive in the sustainability era and adapt to the ever-changing demands of the workforce.
2. **Engage in continuous learning:** Stay ahead of the curve by constantly updating your skills and knowledge. Embrace lifelong learning and take advantage of the vast resources available online, such as courses, webinars, podcasts, and articles.
3. **Expand your network:** Connect with sustainability leaders, experts, and fellow game changers through sustainability-

focused events, conferences, and workshops. Foster meaningful relationships and collaborations that will propel you forward in your career.

4. **Seek mentorship and guidance:** Reach out to experienced professionals in the field of sustainability and related industries to learn from their insights and experiences. Their wisdom can help you navigate the complexities of the sustainability landscape and ensure your success.
5. **Volunteer or collaborate on sustainability projects:** Gain hands-on experience and demonstrate your capabilities by working on real-world sustainability projects. This not only enhances your skills but also showcases your value to potential employers and collaborators.

Taking these steps will allow you to become an influential game changer in the sustainability era and contribute to creating a better tomorrow through your unique skills and values.

The future is now. Are you ready to shape it?

Join the IIEC Sustainability Certification Program today available at https://thebestmba.org and become a game changer.

I hope you enjoyed this journey together, and I would love to hear your thoughts on sustainability and your experience with this program. Feel free to reach out to me via LinkedIn:

https://www.linkedin.com/in/angelberniz/

To a sustainable and prosperous future for all,

Angel Berniz

Made in the USA
Monee, IL
04 November 2026

67384034R00174